Vol. 2
Intelligent Systems Technologies

Computer Aided and Integrated Manufacturing Systems

A 5-Volume Set

Vol. 2
Intelligent Systems Technologies

Computer Aided and Integrated Manufacturing Systems

A 5-Volume Set

Cornelius T Leondes
University of California, Los Angeles, USA

World Scientific
New Jersey • London • Singapore • Hong Kong

Published by

World Scientific Publishing Co. Pte. Ltd.
5 Toh Tuck Link, Singapore 596224
USA office: Suite 202, 1060 Main Street, River Edge, NJ 07661
UK office: 57 Shelton Street, Covent Garden, London WC2H 9HE

British Library Cataloguing-in-Publication Data
A catalogue record for this book is available from the British Library.

COMPUTER AIDED AND INTEGRATED MANUFACTURING SYSTEMS
A 5-Volume Set
Volume 2: Intelligent Systems Technologies

ISBN 981-238-339-5 (Set)
ISBN 981-238-982-2 (Vol. 2)

Desk Editor: Tjan Kwang Wei
Typeset by Stallion Press

Printed by FuIsland Offset Printing (S) Pte Ltd, Singapore

Preface

Intelligent Systems Technology

This 5 volume MRW (Major Reference Work) is entitled "Computer Aided and Integrated Manufacturing Systems". A brief summary description of each of the 5 volumes will be noted in their respective PREFACES. An MRW is normally on a broad subject of major importance on the international scene. Because of the breadth of a major subject area, an MRW will normally consist of an integrated set of distinctly titled and well-integrated volumes each of which occupies a major role in the broad subject of the MRW. MRWs are normally required when a given major subject cannot be adequately treated in a single volume or, for that matter, by a single author or coauthors.

Normally, the individual chapter authors for the respective volumes of an MRW will be among the leading contributors on the international scene in the subject area of their chapter. The great breadth and significance of the subject of this MRW evidently calls for treatment by means of an MRW.

As will be noted later in this preface, the technology and techniques utilized in the methods of computer aided and integrated manufacturing systems have produced and will, no doubt, continue to produce significant annual improvement in productivity — the goods and services produced from each hour of work. In addition, as will be noted later in this preface, the positive economic implications of constant annual improvements in productivity have very positive implications for national economies as, in fact, might be expected.

Before getting into these matters, it is perhaps interesting to briefly touch on Moore's Law for integrated circuits because, while Moore's Law is in an entirely different area, some significant and somewhat interesting parallels can be seen. In 1965, Gordon Moore, cofounder of INTEL made the observation that the number of transistors per square inch on integrated circuits could be expected to double every year for the foreseeable future. In subsequent years, the pace slowed down a bit, but density has doubled approximately every 18 months, and this is the current definition of Moore's Law. Currently experts, including Moore himself, expect Moore's Law to hold for at least another decade and a half. This is hugely impressive with many significant implications in technology and economics on the international scene. With these observations in mind, we now turn our attention to the greatly significant and broad subject area of this MRW.

"The Magic Elixir of Productivity" is the title of a significant editorial which appeared in the *Wall Street Journal.* While the focus in this editorial was on productivity trends in the United States and the significant positive implications for the economy in the United States, the issues addressed apply, in general, to developed economies on the international scene.

Economists split productivity growth into two components: Capital Deepening which refers to expenditures in capital equipment, particularly IT (Information Technology) equipment: and what is called Multifactor Productivity Growth, in which existing resources of capital and labor are utilized more effectively. It is observed by economists that Multifactor Productivity Growth is a better gauge of true productivity. In fact, computer aided and integrated manufacturing systems are, in essence, Multifactor Productivity Growth in the hugely important manufacturing sector of global economics. Finally, in the United States, although there are various estimates by economists on what the annual growth in productivity might be, Chairman of the Federal Reserve Board, Alan Greenspan — the one economist whose opinions actually count, remains an optimist that actual annual productivity gains can be expected to be close to 3% for the next 5 to 10 years. Further, the Treasure Secretary in the President's Cabinet is of the view that the potential for productivity gains in the US economy is higher than we realize. He observes that the penetration of good ideas suggests that we are still at the 20 to 30% level of what is possible.

The economic implications of significant annual growth in productivity are huge. A half-percentage point rise in annual productivity adds $1.2 trillion to the federal budget revenues over a period of 10 years. This means, of course, that an annual growth rate of 2.5 to 3% in productivity over 10 years would generate anywhere from $6 to $7 trillion in federal budget revenues over that time period and, of course, that is hugely significant. Further, the faster productivity rises, the faster wages climb. That is obviously good for workers, but it also means more taxes flowing into social security. This, of course, strengthens the social security program. Further, the annual productivity growth rate is a significant factor in controlling the growth rate of inflation. This continuing annual growth in productivity can be compared with Moore's Law, both with huge implications for the economy.

The respective volumes of this MRW "Computer Aided and Integrated Manufacturing Systems" are entitled:

Volume 1: Computer Techniques
Volume 2: Intelligent Systems Technology
Volume 3: Optimization Methods
Volume 4: Computer Aided Design/Computer Aided Manufacturing (CAD/CAM)
Volume 5: Manufacturing Process

A description of the contents of each of the volumes is included in the PREFACE for that respective volume.

Intelligent Systems Technology is the subject for Volume 2. Knowledge-based systems methods will be utilized in overall manufacturing systems, neural network systems techniques will play an increasingly significant role in manufacturing processes such as the optical inspection of machined parts and collaborative knowledge-based systems techniques will become increasingly significant in optimum product design and manufacturing. Automated process planning will become well integrated part of the very complicated problem of process planning in manufacturing systems. On-line real time monitoring of major manufacturing system elements such as machine tools will become increasingly utilized. Automated visual inspection systems for quality control will become an integral part of manufacturing processes. Internet-based manufacturing systems techniques will become an increasingly significant factor. These and numerous other topics are treated rather comprehensively in Volume 2.

As noted earlier, this MRW (Major Reference Work) on "Computer Aided and Integrated Manufacturing Systems" consists of 5 distinctly titled and well-integrated volumes. It is appropriate to mention that each of the volumes can be utilized individually. The significance and the potential pervasiveness of the very broad subject of this MRW certainly suggests the clear requirement of an MRW for a comprehensive treatment. All the contributors to this MRW are to be highly commended for their splendid contributions that will provide a significant and unique reference source for students, research workers, practitioners, computer scientists and others, as well as institutional libraries on the international scene for years to come.

Intelligent Systems Technology is the subject for Volume 2. Knowledge-based systems methods will be utilized in overall manufacturing systems, neural network systems techniques will play an increasingly significant role in manufacturing processes such as the optical inspection of machined parts, and collaborative knowledge-based systems techniques will become increasingly significant in optimum product design and manufacturing. Automated process planning will become a well integrated part of the very complex problem of automated planning in manufacturing systems. On-line real time monitoring of major manufacturing system elements such as machine tools will become increasingly utilized. Automated visual inspection systems for quality control will become an integral part of manufacturing processes. Internet-based manufacturing systems techniques will become an increasingly significant factor. These and numerous other topics are treated rather comprehensively in Volume 2.

As noted earlier, this MRW (Major Reference Work) on Computer Aided and Integrated Manufacturing Systems consists of 5 distinctly titled and well-integrated volumes. It is appropriate to mention that each of the volumes can be utilized individually. The significance and the potential pervasiveness of the very broad subject of this MRW certainly suggests the clear requirement of an MRW for a comprehensive treatment. All the contributors to this MRW are to be highly commended for their splendid contributions that will provide a significant and unique reference source for students, research workers, practitioners, computer scientists and others, as well as institutional libraries on the international scene for years to come.

Contents

Contents

CHAPTER 1

KNOWLEDGE BASED SYSTEMS TECHNIQUES IN THE INTEGRATION GENERATION AND VISUALIZATION OF ASSEMBLY SEQUENCES IN MANUFACTURING SYSTEMS

XUAN F. ZHA

Manufacturing System Integration Division,
National Institute of Standards and Technology,
Gaithersburg, MD 20899, USA
E-mail: zha@cme.nist.gov

The problem of assembly process planning is critical for the automation and integration of production, due to the combinatorial complexity and the requirement of both flexibility and productivity. This chapter presents an integrated knowledge-based approach and system for automatic generation, evaluation and selection, and visualization of assembly sequences. In this chapter, information and knowledge about a product and its assembly processes is modeled and represented by using integrated object model and generic P/T net formalisms. The comprehensive knowledge-based integration coordinates design and assembly sequence planning in the complex interactions and domain knowledge between the technical and economical aspects. By using the integrated representational model, all feasible assembly sequences are generated by decomposing and reasoning the leveled feasible subassemblies, and represented through Petri net modeling. Both qualitative and quantitative constraints are then used to evaluate each assembly part and operation sequence individually and the entire sequences as well. Based on assemblability analysis and evaluation and predefined task time analysis, estimates are made for the assembly time and cost and operation's difficulty of product when each of these sequences is used. Some quantitative criteria such as assembly time and cost, operation difficulty and part priority index are applied to select the optimal assembly sequence. Finally, a prototype integrated knowledge-based assembly planning system is developed to achieve the integration of generation, evaluation and selection, and visualization of the assembly sequences.

Keywords: Assembly modeling and design; assembly planning; integration; generation; visualization; artificial intelligence; knowledge-based systems.

1. Introduction

Assembly plans, in which parts or subassemblies are put together or operation tasks are executed, can drastically affect the efficiency of the assembly process. For example, a particular sequence may require less fixturing, less changing of tools, having simpler and more reliable operations than others. The field of assembly planning arose to address the issue of how a detailed operation plan could be generated given a high level description of a product to be assembled. Traditionally, the product assembly sequence is planned by an experienced production engineer. The planning of assembly sequences is sometimes a trivial and error-prone task because of the possibility of a large number of potential assembly sequences in a complex assembly, especially in a concurrent flexible assembly engineering environment. The problem of assembly process planning is particularly critical for the automation and integration of production, because of the combinatorial complexity and the need for both flexibility and productivity. Due to the frequent changes of the product design and manufacturing strategies in a time-based competition, there is a growing need to automate the generation and visualization of the assembly sequences. Both cycle time reduction and task parallelism increment require a technique with flexibility, efficiency and parallelism, and suitable to the generation and visualization of assembly plans.

Many research activities have focused on various aspects of assembly sequence planning such as assembly modeling,[19] assembly evaluation,[34] assembly sequence representation and generation. There are algorithms that automatically develop feasible assembly sequences based on geometric constraints. These algorithms, however, have not been integrated into an interactive system that helps the designer and the manufacturing engineer in carrying out assembly planning. The complexity of assembly planning requires the application of artificial intelligence techniques such as knowledge-based systems to represent, reason, and to search for assembly knowledge in designing and developing intelligent assembly planning systems. Part precedence, tool changes, assembly directions, goal positions, possible part grasping zones, and even associated robot motions are some of the considerations in these optimized and intelligent approaches. However, these methodologies do not often fit the needs of a real product assembly in concurrent environment, which involves more complex requirements such as geometric relations, performance measurement and evaluation, resource scheduling, kinematics control, integration of design and planning, etc. The combination of these factors makes the real assembly planning more difficult. Therefore, the existing individual intelligent methods and theories which have been developed for the "block world" or the simple assembly in a specific domain cannot be applied directly to the complex real product assembly systems. The development of integrated intelligent capabilities of computer mediated tools for assembly planning has remained as a challenging research topic.

The objective of this chapter is to present an integrated knowledge-based approach and system for automatic generation and visualization of assembly sequences in an integrated environment. The comprehensive knowledge-based integration coordinates the design and assembly sequence planning in complex interactions and domain knowledge between the technical and economical aspects at the early design stage. The computer-mediated system is based on the knowledge about the design and the designer, and it integrates design with analysis and consultant programs that can apply the knowledge of different relevant domains and then advise the designer on how to improve the product. Graphic portrayal and visualization of designs and supporting information can assist the designer in creating designs, making decisions, solving problems, correcting design errors, and visualizing complex, three-dimensional relationships.

In this chapter, the information and knowledge about a product and its assembly process, e.g. assembly constraints, solid model and CAD database, heuristic rules, etc. is modeled and described by a comprehensive knowledge base. Based on this representation, all feasible assembly sequences of the product are generated by reasoning and decomposing the leveled feasible subassemblies, and then represented through the Petri net modeling. Qualitative strategic constraints are then used to evaluate each assembly part and operation sequence individually, and finally, evaluate the entire sequences as well. The assembly operations are evaluated based on various criteria such as the time and the equipment required. However, this chapter focuses on the analysis of operation's difficulty, which is an important factor and can be translated into time, probability of quality failure, and other significant factors. In addition, this parameter can serve as a useful tool for design and sequence evaluation. In order to obtain a good assembly sequence, some quantitative criteria such as assembly time and cost, operation difficulty and part priority index are applied to select the optimal assembly sequence. Based on assemblability analysis and evaluation, and predefined task time analysis, estimates are made of the assembly time and cost and operation's difficulty of the product when each of these sequences is used. Finally, an integrated knowledge-based assembly planning system has been developed to achieve the integration of the generation, the selection and evaluation, and the visualization of the assembly sequences.

2. Review of Related Work

From the literature, many attempts have been made to carry out various aspects of assembly design and planning, such as the development of computer-aided design and planning systems, the evaluation of assembly design and planning, and the strategies to facilitate assembly process.[2–5,9,10,90] By definition, the field of assembly and task planning can be broken down into three major areas: the integration of design and manufacturing as it pertains to assembly planning; general off-line assembly and task planning; on-line planning, execution, and reaction.[68] Assembly planning is generally considered to be the process in determining a set of instructions for mechanically assembling a product from a set of sub-components. Each instruction will usually specify that a sub-component be added onto the partially-completed assembly in a particular way, such as a nut screwed onto a bolt or a lid press-fitted onto a box-top. Research in assembly planning was initially aimed at assisting process planning to reduce delays between design and manufacturing and to produce better plans.[29] Most published assembly planning contributions focus on the modeling of assembly process, i.e. describing the geometric configurations of the assembly which is constructed by single parts, as well as the order among the parts. In recent years, interest has shifted towards generating assembly sequences to evaluate assembly designs and to create products that are easier to manufacture. Automated geometric reasoning and computational efficiency of assembly planning have become more critical. The geometric approach to assembly planning originated in robotics have been reported in Refs. 16 (AUTOPASS), 25 (LAMA), and 26. It is more limited in scope than traditional artificial intelligence (AI) planning, and focuses specifically on issues raised by the manipulation of physical objects. This has led to work in basic path planning, motion planning with uncertainty, manipulation planning with movable objects, and grasp planning.[27] However, assembly planning presents a complex problem for general motion planning. A simpler subproblem known as assembly sequence planning, or simply assembly sequencing[7,8] was subsequently evolved, where only the geometric constraints arising from the assembly itself are considered.

Several planning methodologies and techniques have been proposed in the literature, using an approach limited to specific product topologies and structures. Some of them are more suitable for dedicated automated assembly and assembly lines rather than for flexible automated assembly, flexible assembly systems and assembly job shops. The early assembly sequencers were mainly sequence editors with geometric reasoning supplied interactively by a human.[28,29] Automated geometric reasoning capability was subsequently developed.[11,12,17–19,23,30–32] These "generate-and-test" assembly sequencers, are equipped to guess candidate sequences and generate questions to check their feasibility by the geometric reasoning modules. They tend to generate repetitive geometric computations. Mechanisms for saving and reusing previous computations, such as the "precedence expressions",[23] have been proposed, but have had limited success. In practice, the generate-and-test paradigm is relatively inefficient as its processing time increases exponentially with the number of parts and is applicable only for assemblies with few parts. The non-directional block graph (NDBG) approach proposed by Wilson[24] and Romney *et al.*[33] circumvents this combinatorial trap. Although there is a combinatorial set of potential part interactions, NDBG represents them in polynomial space, allowing valid operations and assembly sequences to be directly generated in polynomial time.

An assembly can have many different feasible assembly sequences. Due to the difficulty to represent each sequence individually, it is necessary to design a method to represent and visualize all the sequences in an efficient and compact manner. There have been three main approaches to the representation of assemblies, i.e. language-based representation, graph-based representation, and advanced data structure representation, with three different underlying goals.[41] The graph-based approach representation is more general and usually extracts data from more information sources such as a CAD database, or from information supplied by the user. Its forms are numerous,[41] and they include directed graphs, AND/OR graphs,[7] Petri net graph,[49,91] connectivity graphs,[44] hierarchical partial order graphs,[45] liaison diagrams,[29] precedence diagram,[13,46] assembly constraint graphs,[10,47] interference graphs,[48] and knowledge assembly liaison graph (KALG).[88,91]

In the last decade, research efforts have brought about the adaptation of advanced problem-solving methodologies for the solution of long-standing problems in assembly process planning. Inevitably, one of the most common causes of unsuccessful attempts at the development of decisive CAAP systems is a failure in the communication that is necessary between the human planner and the problem-solving procedures.[89] Two approaches have been taken to solve planning problems:[63] understanding and solving the general problem such that the planning system can be expected to work for a reasonably large variety of application domains; and the use of domain specific heuristics to control the planner's operation. The bulk of AI research can be found in the former area of domain-independent research, mainly due to the fact that it is a more difficult problem and its solution is still elusive. In the latter area, practical systems are successfully being developed and are making their way from the research labs to everyday use. The most widely used method for optimum and intelligent assembly planning is to represent the assembly sequence by AND/OR graphs and then use heuristic search methods in AI such as depth-first search methods, breadth-first search methods and AO* algorithms to obtain the optimal assembly sequence.[92] Another method is to use the Petri net based algorithms.[49,91,92] The majority of these successful domain-dependent systems are implemented using a rule-based approach (i.e. expert system) which demonstrates the utility of this new-found tool.

Knowledge-based and computational-intelligence-based approaches such as expert systems, fuzzy logic, and neural networks have been employed to encode assembly planning heuristics into rules or computation models and to generate and optimize assembly plans.[15,41,56–58] These formulated rules are then applied recursively to decompose a product assembly into its constituent parts or subassemblies. Attempts have also been made to translate the constraints imposed by the precedence relationships among assembly tasks into predicate calculus or rules that can be easily employed to disassemble a product assembly. Others have applied production rules to increase the efficiency of the planning process. The relationships among the parts within an assembly may also be modeled with liaison graphs or relational models. By employing graph theory and AI techniques such as heuristic search, the graph is successively decomposed into sub-graphs for the representation of subassemblies or individual parts. All possible sequences can therefore be generated. Assembly sequences that conform to certain criteria are then selected for the generation of feasible assembly plans. Ben-Arieh[57] used a fuzzy-set-based method to evaluate the degree of difficulty of each assembly operation and then select a "best" sequence of assembly operations. Hong and Cho[58] proposed a neural-network-based computational scheme to generate optimized robotic assembly sequences for an assembly product. The hybrid intelligent system used a neural network with functional link nets and an expert system. Based on the assembly constraints inferred and the assembly costs obtained from the expert system, the evolution equations of the network were derived, and an optimal assembly sequence was obtained from the evolution of the network. Chen and Yan[62] also applied neural network computing techniques in the case associative assembly planning system (CAAPS) by integrating memory organization and neural networks computing techniques. The CAAPS system provides an environment with high-level features for synthesizing the assembly rapidly. At all stages of the design process, the user can consult the case associative memory (CAM) for prior experience of similar assembly. It can retrieve past knowledge based on design intentions, part names, and connection types among the parts, and remember the assembly cases on the basis of internal similarity between cases. The precedence knowledge and constraints of operations are stored as the contents of cases.

The other important development offers a more integrated approach to assembly planning by linking aspects of product design, design for assembly, assembly planning and production layout within a manufacturing enterprise in just one procedure.[36–43] All of them require some interactive inputs, which currently seems to be the only way of generating practical results.[50–61,65–70] Some generic-planning-based software in AI such as STRIPS and specific computer-aided assembly process planning software such as SAAT,[33] Archimedes 3.0,[20] MIT-CAAP[9] CMU-CAAP,[71] WSU-CAAP[96] can be applied to assembly planning. Incorporated into a larger CAD tool (e.g. ProEngineer), these systems (e.g. SAAT) can generate and evaluate the geometric assembly sequences of complex products from 20 to 40 parts and from 500 to 1500 faces. Thus they provide immediate feedback to a team of product designers regarding the complexity of assembling the product being designed. The developers of SAAT are now working on extending to more sophisticated cases both these results and the underlying theory. Archimedes 3.0[20] is an interactive assembly planning system. A fast core planner enables a highly productive plan-view-constrain-plan cycle. The user asks for a plan, simulates it in 3D, adds the known process constraints violated by the plan, and then iterates until a satisfactory plan is found. Interactivity is critical to effective assembly planning even for moderately complex products, since many process constraints are difficult to identify until they are violated, and an impractically large number of plans result when process constraints are not considered. The program composes of a central search engine, together with the surrounding modules that apply constraints due to part-part collisions, the subassembly connectivity, the tool and grasping requirements, and the user-defined process constraints. Special attention is paid to the efficiency in every module. However, the most comprehensive one is the project on integrated design and assembly planning (IDAP),[5,46] which is a very long project with large scale interaction. In the IDAP system, a model of operation network (OPNET) is constructed to reflect the constraints intrinsic in the product itself. The operation network graph represents the relationships among tasks and sub-tasks for the design-for-assembly evaluation and the assembly process planning. The interactive operation network editor was developed to integrate the procedures for network generation, modification, queries and evaluation under a uniform graphical man-machine interface.[59] A systematic approach and a computer aided multi-agent system for the concurrent integrated product design and assembly planning (CIDAP) were developed by author and applied successful in lighting products.[94,95]

3. Assembly Modeling and Representation

In designing a product and its related processes, all the product information should be organized and represented as product models within a computer. The effectiveness of a design and planning system relies heavily on the input of the product model representation. The majority of the models used in this domain are simple approximations of the real workpiece and they are often dedicated to special algorithms. A new integrated object model that is particularly useful for model-based assembly design and planning is presented in this section. The integrated object model described here provides a more accurate and more flexible representation of workpieces. It consists of two parts, one describing the geometry and topology, and the other describing the technological properties of an object.

3.1. *Notations and assumptions*

3.1.1. *Notations*

(1) Parts of an assembly are represented as $p_1, p_2, \ldots, p_n$;
(2) A subassembly is represented by a list of parts such as $[p_1, p_2, p_3, p_4]$;
(3) A subassembly operation is represented by a combination of two subassemblies such as $[[p_1], [p_2, p_3, p_4]]$, $[[p_1, p_2], [p_3, p_4]]$, etc. The list is read from left to right, i.e. the subassembly operation direction or sequence is from left to right, for example, $[p_1] \rightarrow [p_2, p_3, p_4]$ and $[p_1, p_2] \rightarrow [p_3, p_4]$;
(4) An assembly sequence is represented implicitly by nested lists of parts. For example, $[[[[p_1, p_2]\ [p_3]], [p_4]], [p_5, p_6]]$ represents the following sequence of assembly operations: $p_1 \rightarrow p_2 \rightarrow p_3 \rightarrow p_4 \rightarrow (p_5 \rightarrow p_6)$.

3.1.2. *Assumptions*

In order to generate all the feasible assembly sequences systematically, the following guidelines are used.

(1) For a disassembleable product, the assembly sequence is the reverse of the disassembly sequence if there is no destructive operation in the disassembly.
(2) Each part is a solid rigid object, that is, its shape remains unchangeable. Parts are interconnected whenever they have one or more compatible surface in contact.
(3) Two types of fasteners may be used by an assembly, i.e. screws and nuts and bolts.
(4) The assembly is built sequentially with a single component added one at a time.

(5) All components are assembled directly to their final positions with a single linear motion. Actions such as "insert and twist" are not allowed.
(6) An assembly operation once completed would remain unchanged at all assembly stages.

3.2. *Generic P/T nets*

From the graph theory,[83] a generic graph can be descried as a two tuple $G = G(V, E)$, where V is the set of nodes and E is the set of connecting arcs linking the nodes. If each arc in a graph has a direction, then the graph is a directed graph. In this chapter, we categorize a node into two classes: place (P) node and transition (T) node. Then a place-transition (P/T) net graph model (Fig. 1(a)) can be defined as: $\mathrm{PTN} = \{\mathrm{P}, \mathrm{T}, \mathrm{F}, \mathrm{W}\}$, where $\mathrm{P} = (p_1, p_2, \ldots, p_m)$ is a place node set; $\mathrm{T} = (t_1, t_2, \ldots, t_n)$ is a transition nodel set; and F is an arcs set linking the place node and transition node, and it has the characteristics: $\mathrm{P} \cap \mathrm{T} = \phi$, $\mathrm{F} \subseteq (\mathrm{P} \times \mathrm{T}) \cup (\mathrm{T} \times \mathrm{P})$, and $\mathrm{P} \cup \mathrm{T} \neq \phi$; and W: $\mathrm{F} \rightarrow \{0, 1\}$ is an association weight function on arcs, $\forall f \in \mathrm{F}$, $\mathrm{W}(f) = w_i$, w_i is the weight of arc f. Similarly, a place-transition net graph is a directed place-transition (P/T) net graph (Fig. 1(b)) if each arc in the graph has a direction.

On the other hand, as defined in Refs. 55 & 76, Petri nets generally consist of places and transitions, which are linked to each other by arcs. They can be described as bipartite directed graphs whose nodes are a set of places associated with a set of transitions. Therefore, a Petri net graph is in fact a directed place-transition net graph. If the net activities are based on a vision of tokens moving around an abstract network, in which tokens are conceptual entities that model the objects and appear as small solid dots moving in a real network, a marked Petri net (Fig. 1(c)) is formally defined as a 5-tuple $\mathrm{PN} = (\mathrm{PTN}, \mathrm{M_0}) = (\mathrm{P}, \mathrm{T}, \mathrm{F}, \mathrm{W}, \mathrm{M_0})$, where PTN is a directed P/T net; $\mathrm{P}, \mathrm{T}, \mathrm{W}, \mathrm{F}$ are the same as the above definitions, and $\mathrm{M_0}$: $\mathrm{P} \rightarrow \{0, 1, 2, \ldots\}$ is the initial marking. The Petri net graph is a graphic representation of the Petri net structure and it visualizes the reasoning rules. Fundamental techniques for the analysis of Petri nets are state space construction, matrix-equation approach, and reduction or decomposition techniques.

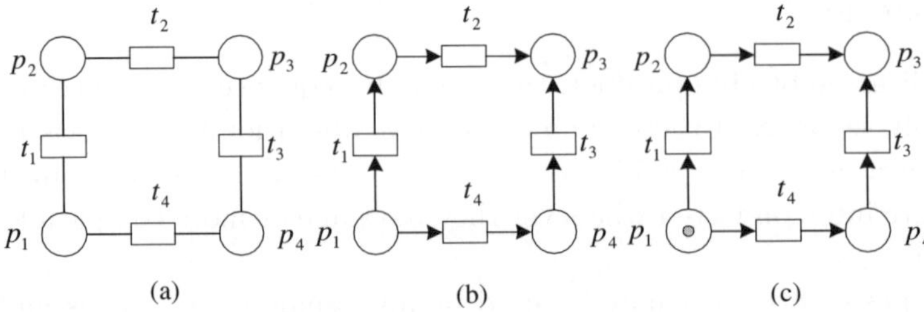

Fig. 1. Generic P/T nets: (a) place-transition net, (b) directed place-transition net or Petri net, (c) marked Petri net.

The construction of a state-space representation called the reachability graph by enumeration allows for the computation of all properties. However its usefulness is limited by the state-space explosion which often occurs even in seemingly simple models. In spite of this capability, feasibility and efficiency considerations motivate the use of the algebraic approach, as well as the reduction or decomposition techniques whenever possible.[76]

For a Petri net with m places and n transitions, the incidence matrix $A = [a_{ij}]$ is defined as an $m \times n$ matrix of integers whose generic entry is given by $a_{ij} = w(j,i) - w(i,j)$, where $w(j,i)$ is the weight of the arc from transition j to its output place i. Thus the entry a_{ij} of the incidence matrix A represents the negative or positive change in the number of tokens in place i due to the firing of transition j. The integer positive vector y solutions of the homogeneous equation $A^T \cdot y = 0$ are called the P-semiflows of the Petri net. The integer positive vector x solutions of the homogeneous equation $A \cdot x = 0$ are called the T-semiflows of the Petri net. P- and T-semiflows are computed from the incidence matrix and do not depend on the initial marking.[76] Each P-semiflow identifies an invariant relation stating that the sum of tokens in all places, weighted by y, is constant for any reachable marking and is equal to $\mathrm{M}_0 \cdot y$ for any initial marking M_0. This invariant relation is called a P-invariant. Each T-semiflow identifies an invariant relation (T-invariant) stating that from marking M the same marking is achieved by firing any transition sequence whose firing count vector is given by the T-semiflow x, provided that such a sequence can actually be fired from M. Other concepts useful for analyzing Petri nets are the notions of deadlocks and traps. A deadlock is a subset of places $\mathrm{P}_s \subseteq \mathrm{P}$ such that the set of its input transitions is a subset of its output transitions. A trap is a subset of places $\mathrm{P}_t \subseteq \mathrm{P}$ such that the set of its output transitions is a subset of its input transitions. The total number of tokens in a deadlock cannot increase, and the total number of tokens in a trap cannot decrease. A P-semiflow is both a deadlock and a trap.

3.3. *Object oriented knowledge representation*

The integrated object model is, in fact, an attempt to set up a knowledge framework in such a way that it becomes possible to process various types of knowledge in a top-down design process. Information processing in machine design is inherently model-based, since the design object is structural in type. Therefore, object-oriented programming languages for knowledge representation are desirable. Object-oriented programming technique can allow designers to look at a design problem as a collection of objects or sub-problems linked together by rules. Thus it provides the designers with an expressive power to represent complex problems or information in an effective manner. If a designer can break a design problem in the form of well defined clearly manipulatable chunks with their own self-containing information which are interrelated through a series of rules and constraints, then these problems will be led themselves well to object-oriented programming applications and then conveniently solved. This chapter concentrates on introducing integrated knowledge representations related to design and planning. It deals mainly with declarative representation, production rules and object oriented concepts. Procedural representation using conventional languages such as C will not be emphasized. Design and planning processes and activities, especially for the more complex type can be fulfilled by integrating knowledge in its multiple forms, levels and functions. The integration process is very challenging as the overall effect may be greater than the sum of its parts. The integrated knowledge can solve problems which cannot be attained by the individual knowledge alone.

The object-oriented knowledge representation is based on a mixed representative method and the object-oriented programming (OOP) techniques. The basic structure of this representation is described as a unit. The class of an object and its instances are described by the unit structure. An object-oriented unit is composed of four types of slots, which are the relation slot, the attribute slot, the method slot and the rule slot. The relation slot is used for describing the static relations among objects or problems. With the help of the relation slot, according to the relation of classification, the design object can be described as a hierarchical structure. The knowledge in superclass can be shared by its classes and subclasses. The messages that control the design process can be sent among all instances of objects. In addition, if needed, other kind of relation slots can be defined, such as the resolution, the position and the assembly, etc. These slots create the foundation for describing graphs in design. The hierarchical structure of object oriented knowledge representation is exemplified and illustrated in Fig. 2. The attribute slot is used for describing the static attributes of design object, such as the tooth number of gear, its module number, material, etc. The method slot is used for storing the methods of design, sending messages, performing procedural control and numerical calculations. The rule slot is used for storing sets of productions rules. The production rules can be classified according to the differences among objects being treated and stored respectively in rule slots in the form of slot value.

3.4. *Representation of MSAs: Integrated object model*

The integrated object model for mechanical systems and assemblies (MSAs) is defined by a hierarchy of structure, geometry and feature. A structural model is a component-connector multi-graph. The corresponding multi-graph is used for uniformly describing their causal relations. The model proposed is a uniform description for both assembly level and feature-based single part level. The model allows designers to deal with incomplete, imprecise knowledge and uncertainty with the help of fuzzy logic.

3.4.1. *Hierarchical structure description*

For an assembly composed of parts or components and connectors (joints), or a single part composed of physical features, the different levels of assembly actually

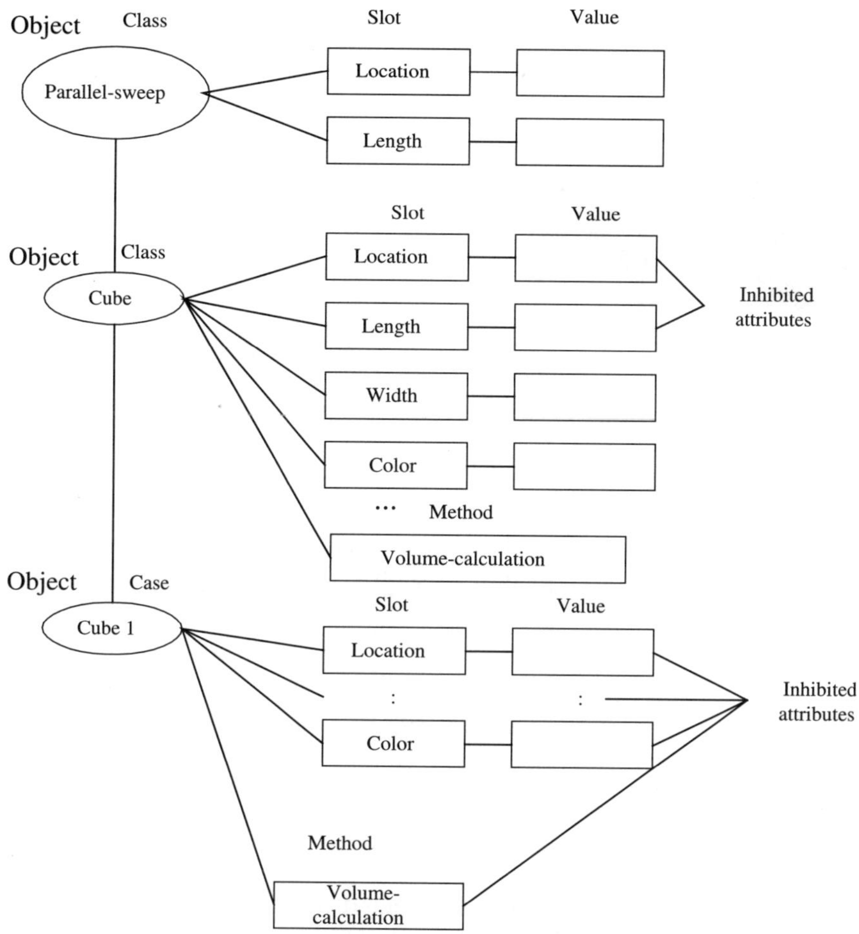

Fig. 2. Object oriented knowledge representation.

form a hierarchy which utilizes the relationships between different parts of the assembly.[67] The "place-transition" model is used to represent the mechanical systems and assemblies, in which each part is represented as a place and each connector is represented as a transition. A mechanical system (assembly) is therefore a hierarchical P/T net (called the *Assembly_Model*) and a subsystem or subassembly is therefore a sub P/T net. Using modular representation, a sub P/T net can be described as either a macro place or a macro transition. This is dependent mainly on its function as a component, a joint or a connector. Token data abstraction and dynamic distribution can be used for knowledge representation in describing the structure and system state changes. The definition of machine structure using P/T net is as follows. Various relations between place nodes and transition nodes in a hierarchical P/T net can be clarified with reference to Fig. 3.

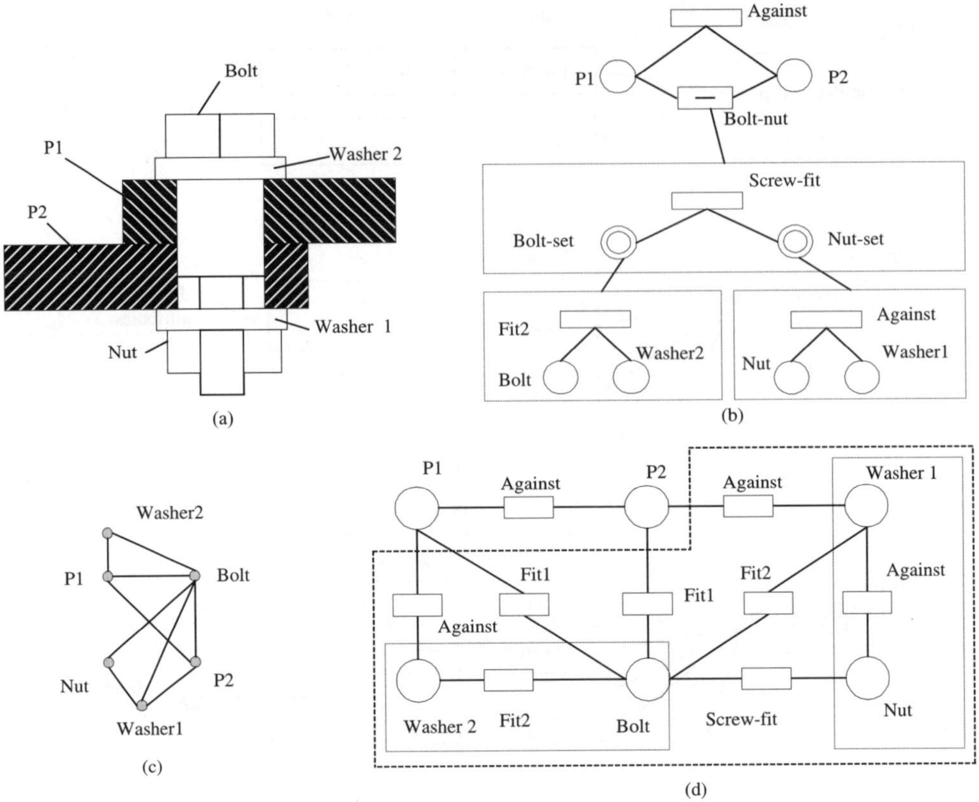

Fig. 3. Graphical relations of place nodes and transition nodes.

Therefore, an assembly structure represented by place-transition model can be defined as: $\mathrm{S-PTN} = \{\mathrm{P}, \mathrm{T}, \mathrm{F}, \mathrm{W}\}$, where $\mathrm{P} = (p_1, p_2, \ldots, p_m)$ is a place set which represents objects consisting of components; $\mathrm{T} = (t_1, t_2, \ldots, t_n)$, is a transition set which represents joints; and F is an arcs set which links between components and joints; and W is a weights set of the arcs. For a hierarchical P/T model, as shown in Fig. 4, the definitions are presented as follows:

(1) A structure on the top level is only a P/T graph, S_0 with one macro place or macro transition node;

(2) A structure on ith level $(i = 1, \ldots, L)$ S_i is a graph: $\mathrm{S}_i = \{\mathrm{P}_i, \mathrm{T}_i, \mathrm{F}_i, \mathrm{W}_i\}$, where P_i is a set of places denoting components c_{ij} or subassemblies sub_{ik}, i.e. $\mathrm{P}_i = \{c_{ij}, sub_{ik}\}$, $(j = 1, \ldots, h_i,\ k = 1, \ldots, x_i)$; T_i is a set of transitions denoting either joints J_{is} or connectors l_{it}, i.e. $\mathrm{T}_i = \{J_{is}, l_{it}\}$ $(s = 1, \ldots, m_i,\ t = 1, \ldots, n_i)$; A_i is a set of arcs linking J_{is} or l_{it} and c_{ij} or sub_{ik}, i.e. $\mathrm{F}_i = \{f_{isj}, f_{isk}, f_{itj}, f_{itk}\}$; W_i is a set of weights of arcs, i.e. $\mathrm{W}_i = \{w_{isj}, w_{isk}, w_{itj}, w_{itk}\}$. The structure can be also expressed by a collection of unconnected graphs $s_{i,a}$ $(a = 1, \ldots, y)$, i.e. $\mathrm{S}_i = \{s_{i,a}\} = \{\{p_i, t_i, f_i, w_i\}_a\}$, $p_i \in \mathrm{P_i}$, $t_i \in \mathrm{T_i}$, $f_i \in \mathrm{F_i}$, and $w_i \in \mathrm{W_i}$.

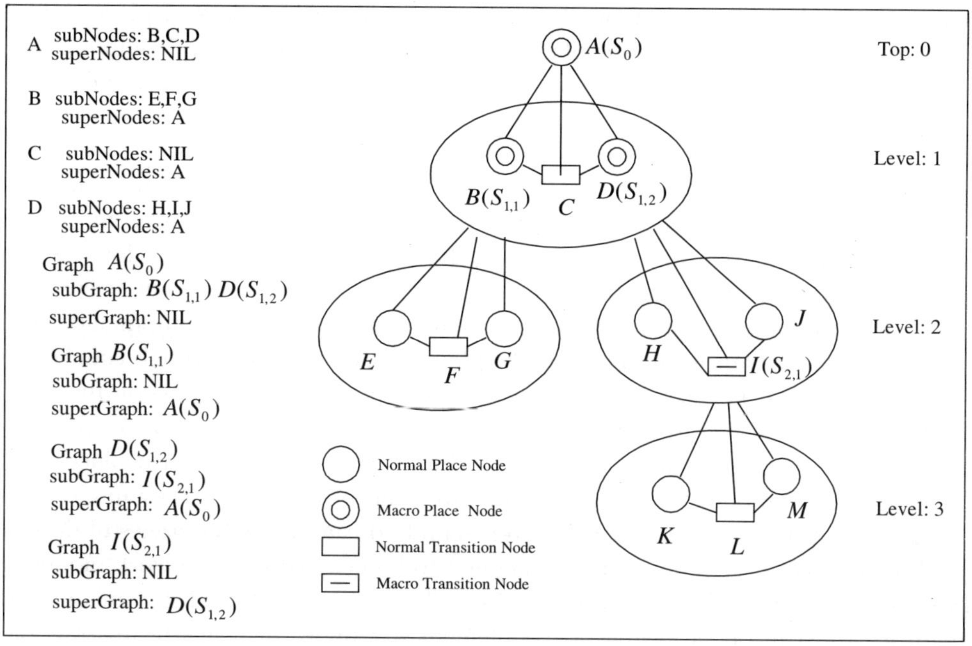

Fig. 4. Hierarchical P/T model.

(3) A place or transition (p_i or t_i) or graph $s_{i,a}$ may be associated with another graph $s_{i+1,b} = \{\{p_{i+1}, t_{i+1}, f_{i+1}, w_{i+1}\}_b\}$. Such a graph is a subgraph of the graph $s_{i,a}$; vice versa, $s_{i,a}$ is a super graph of the graph $s_{i+1,b}$. The place or transition is termed macro place or macro transition, either a subassembly place or a connector transition.

(4) Assembly incidence matrix: Suppose an assembly P/T net that has n transitions and m places. Its incidence matrix is $C = [C_{ij}]$ $(1 \le i \le n,\ 1 \le j \le m)$. We define this incidence matrix as assembly incidence matrix. Every row of C represents a transition or a macro transition; every column represents a place or a macro place, $C_{ij} = w(t_i, p_j) - w(p_j, t_i)$. For the bolt-nut fastening assembly (Fig. 3), the assembly incidence matrix can be described as follows:

$\mathcal{AIM}(C) =$

	$\mathbf{P}_1(p_1)$	$\mathbf{P}_2(p_2)$	**washer1**(p_3)	**washer2**(p_4)	**bolt**(p_5)	**nut**(p_6)
against1$(\mathbf{t_1})$	1.0	1.0	0	0	0	0
against2$(\mathbf{t_2})$	0	1.0	1.0	0	0	0
against3$(\mathbf{t_3})$	1.0	0	1.0	0	0	0
against4$(\mathbf{t_4})$	0	0	1.0	0	0	1.0
$\mathbf{fit1-1(t_5)}$	1.0	0	0	0	1.0	0
$\mathbf{fit1-2(t_6)}$	0	1.0	0	0	1.0	0
$\mathbf{fit2-1(t_7)}$	0	0	1.0	0	1.0	0
$\mathbf{fit2-2(t_8)}$	0	0	0	1.0	1.0	0
$\mathbf{screw-fit(t_9)}$	0	0	0	0	1.0	1.0

Thus any machine consist of two basic classes of objects: a class of places and a class of transitions. Basically, a place is a named concrete object (component or subassembly) which performs desired functions of machine in possible behaviors through the connection with the other places. Examples are a unit with no detail or physical entity, e.g. a step shaft, a case or a sub-assembly. A transition is a named abstract object corresponding to joints, constraints, or operations and functions between two components. It does not provide constraints to the relevant components until the components are mated together. Its properties could also allow for the possibility that the properties take some values in the form of a fuzzy set over a base range. P/T net can model mechanical causal relations between components. The main purposes of transitions are to make places work normally through connecting these components. For example, a transition with a motion transmission function, might become a gear pair; a transition with a fixing function might be a collection of geometric mating surfaces such as a cylinder and shoulder. Since places and transitions for components and connectors are conceptually fuzzy, they might form a fuzzy P/T net which represents a sub-assembly during the later stages of design.

A multi-level P/T net could be imagined to be generated by network modeling from top to bottom. The example "bolt-nut" structure in Fig. 3(a) can be used again for illustration in Fig. 5. To implement a number of levels of abstraction, a usual network model in macro place and transition is split into several embedded blocks corresponding to the designer's thinking patterns on various levels, or say, the think blocks. On the top level, designer's goal of fixing plates P1 and P2 is expressed as putting a connector between the components P1 and P2. Thus a labeled rectangular block (macro transition) which is located between P1 and P2 replaces the thinking block on the first level after the first abstraction. A further splitting of the second thinking block produces the "bolt-set" and the "nut-set" which are logical components. Two labeled circles replace the thinking block on the second level after the second abstraction. The replacement of a part of the original network model by a logic node is a type of abstraction, through which nodes in the network model are pulled down one level as shown in Fig. 5. In this way, a multi-level Petri net graph with increasing structural information details as in Fig. 5 could be naturally created during the in-progress design. The significant improvements from the use of a multi-level P/T net are carried out in two steps: Firstly, the nodes of a multi-level P/T net graph are functionally divided into places (components) and transitions (joints). Then the normal nodes which denote atomic components are graphically distinguished from macro nodes which are associated with another level P/T net graph. Such a distinction reflects various levels of abstractions at different stages of a design process. One of the advantages of using a high abstraction of connectors is that the feature-based modeling for single-piece parts becomes a natural extension of assembly modeling. On the lowest level, the connectors (transitions) are the features of single components that mate and structure these components.

Using object-oriented representation, the attributes and functions of the *Assembly_Model* are represented as follows. The *Assembly_Model* class carries the

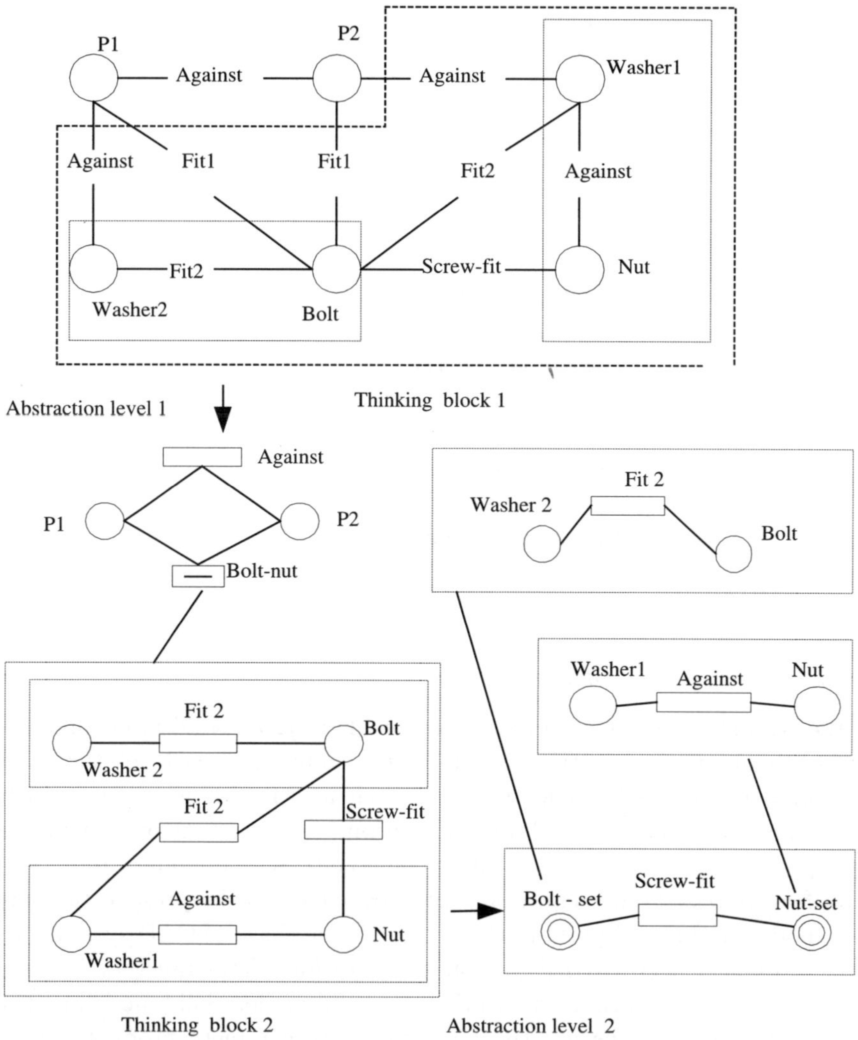

Fig. 5. A multi-level P/T net graph for the "bolt-nut" structure.

is-part-of relationship of a mechanical system and its components. The attributes and methods (functions) of the *Assembly_Model* are defined to help the designer construct the structure of the mechanical system. The editing functions allow the designer to create specific system configurations. When a designer creates a new system, the system configuration (or decomposition) is based on some special purposes from the designer's viewpoint. Furthermore, based on different application considerations, the designer can edit the configuration or evolve it through experiments to conceive, as a meaningful unit, an analysis that evaluates some aspects of the performance of the system or subsystem.

```
Class Assembly_Model
{
Attributes:
            ID
            Name
            Set of Assembly_Models (Nil or Composite IDs)
            Set of transitions (Joints) (Nil or Joint IDs)
Methods:
            Create Assembly_Model
            Add place (Part)
            Erase Place (Part)
            Add transition (Joint)
            Erase transition (Joint)
            Add Assembly_Model
            Erase Assembly_Model
            Display place (Part)
}
```

3.4.2. *Geometries in assembly model*

Since the research interest is to reason about 3D objects, solid modeling has been chosen to create the geometric models of the individual parts of the assembly. Other common representational methods, such as wire-frame systems, instances and parameterized shapes, cell decomposition, spatial-occupancy enumeration, sweep representation, constructive solid geometry (CSG), boundary representation (B-rep), feature-based representation, and knowledge-based representation, are also used to model mechanical parts in the geometric database of CAD systems. Recent research work has been carried out on the addition of tolerance of information, whether dimensional or geometric, to the part solid model. Solid modeling, features and attribute relationships are the basis for more complete product definition.[6] In addition to rigorously defining geometry and topology of individual parts and joints above, product assemblies are defined through solids primitive modeling by defining the:

(1) Instances or occurrences of each part in a hierarchical manner;
(2) The relative location of each instance or occurrence of the part in terms of the part's x, y and z coordinates relative to the assembly's base or reference point x, y and z coordinates;
(3) For each instance or occurrence of a part, the part's orientation is in relation to the assembly's orientation;
(4) Vectors or axes of rotation and translation to describe the movement of parts within assemblies.

This approach can yield a complete definition of the product's geometry and topology at any level in the product structure. Many assembly relationships (e.g. topological liaison, geometric liaison) and constraints (e.g. geometric constraints, and partial precedence constraints) discussed in assembly planning below are extracted or reasoned out from the defined assembly geometric model.

3.4.3. *Features and semantics in assembly model*

In the integrated object model, the form feature, precision features, and assembly features are organized in the mechanical system's hierarchical structure. Form features and precision features are embedded in the part object, while assembly features are carried by the joint object. Form features are the geometric features which are designated to represent the part's shapes. A form feature is carried by the geometric representation of the part. Precision features include tolerances and surface texture which are also grouped under the same composite attribute (geometric representation). Assembly features are particular form features that affect assembly operations.

Each form feature has certain precision features associated with it. For example, a slot (form feature) has dimensions such as height, width, and length; each

Geometric Entity[42]**:**

{ *its super class-object*
Sub-class-part and feature
Instance variables
Name: the unique identifier
Type: the sort of the object

Part:
its super class-Geometric Entity
Sub-class-feature
Instance variables
Component: features a part holds
Neighbor: related part's holds
Related component: features having relations with other parts
n-relation: related part along a specific direction, n, and the position of the related feature, where n=+X,+-Y,+-Z
n-list: a list of features whose normal is n with the order from nearest to farthest along n, where n=+X,+-Y,+-Z

Feature: its super class-GeometricEntity
Instance variables
Location: position (x,y,z) and orientation (nX,nY,nZ) of a feature
Relation: related part name

}

dimension has tolerance (e.g. positional tolerance, straightness, or perpendicularity to some datum) and surface finish (e.g. lay direction, average surface roughness). When parts mate together, both the parts' form features and precision features govern the assembly operations. The parts' form feature directly affect the joining conditions. For instance, a hole or pin indicates the fit condition, a threaded stud or threaded hole suggest that a torque operation is needed. Obviously, the precision features of the mating parts also affect the quality and manufacturing processes of the assembly.

It has been shown that feature-based product models for assembly can help considerably in both assembly modeling and planning, on the one hand by integrating single-part and assembly modeling, and on the other hand by integrating modeling and planning. For modeling and planning of both single parts and assemblies, an integrated object-oriented product model has been developed. For specific assembly-related information, assembly features are used. Handling features contain information for handling components, and the connection features contain information on connections between components. Therefore, for a product and its assembly process, both the part and feature level information are required. This can be summarized as[42]: part (name, type, class, material, heat-treatment) $\leftrightarrow$ feature (name, type, parameters, locations, tolerances, relations, surface-finish). As discussed before, one of the advantages of using a high abstraction of connectors or joints is that feature-based modeling for single-piece parts becomes a natural extension of assembly modeling. On the lowest level, connectors or joints can be considered as the features of single components that mate and structure components. In Fig. 6, the feature model for a step shaft in an assembly consist of the shaft and a set of connectors: keyFit, Spline Fit and several cases of FixFit, all of which are the usual features. Thus a unified description of a feature-based model of both an assembly and single piece components will be obtained through this data abstraction of components and connectors on various levels from function-behavior-structure-based conceptual modeling, geometric modeling to feature-based design. The first part is a shaft whose design and feature-level description is shown in Fig. 7.

3.5. *Representation for assembly planning*

As discussed above, information about joints and parts of a mechanical system, as represented by the "place-transition" model in the global product definition, can be used for assembly process planning. This is because parts are the elementary components for making an assembly and joints carry the connectivity information of parts, which points to the assembly features of parts. The assembly features of joints as defined in the global definition indicate how parts are mated or jointed together. Although both the part and joint global definition contain necessary information for the assembly process, additional information is required for assembly planning.

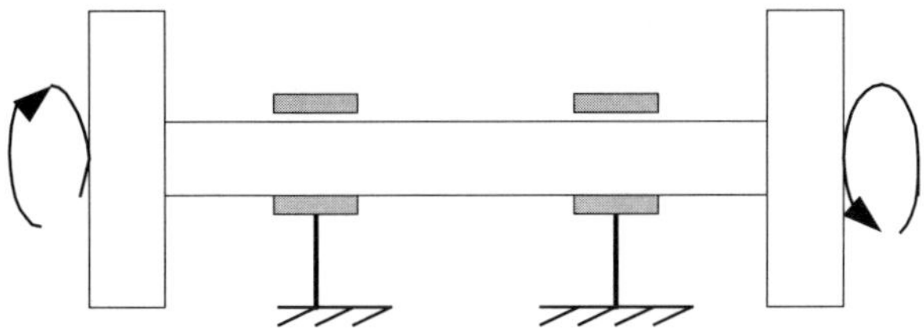

(a) Conceptual level

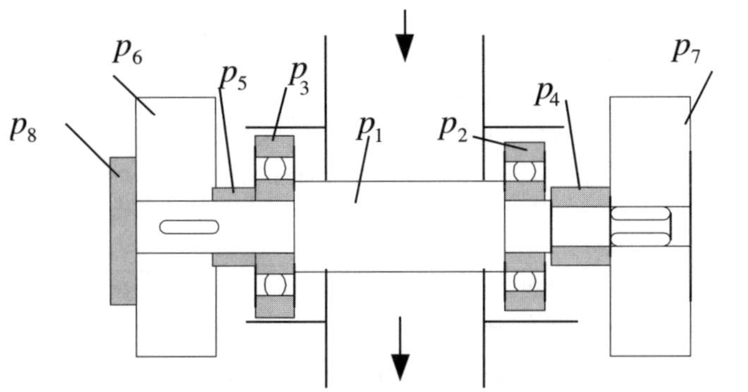

(b) Layout level

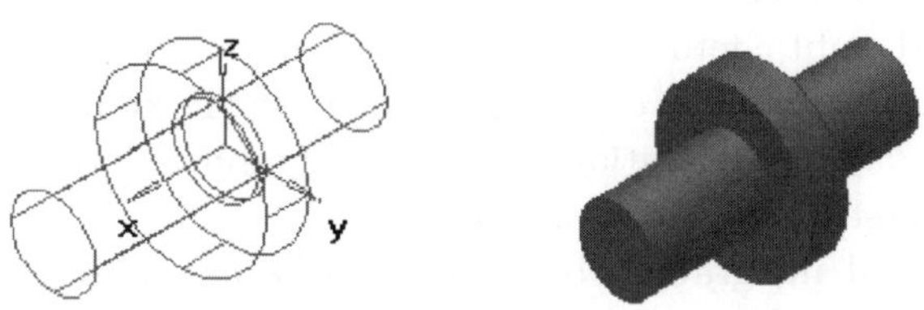

(c) Geometric modeling level

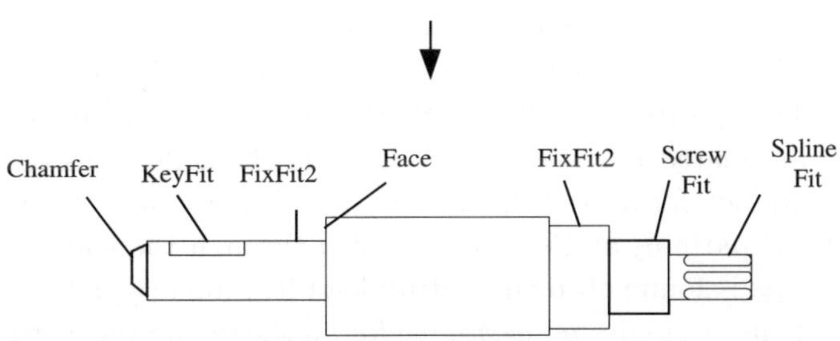

(d) Feature level

Fig. 6. Data abstraction at different assembly levels.

3.5.1. *Relational models for assemblies*

(1) Topological liaison model

Two parts are said to be in contact if they are constrained to touch along a surface, line or point. The liaison relationship between the two parts can be implemented

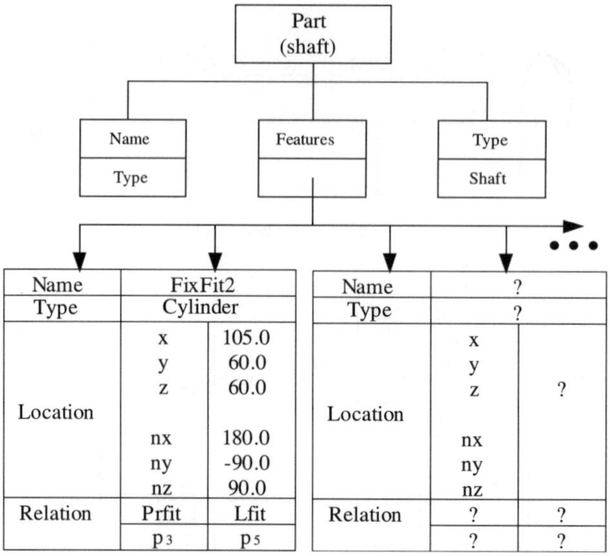

Fig. 7. Feature-based shaft representation.

by the total relative constraints which can be extracted from the assembly models and the classified as fit contact, plane contact, and meshing contact. The assembly liaison relations of a product can be described by the knowledge assembly liaison graph ($KALG$).[91] The generation of a liaison graph is based on the information retrieved from the feature model of a product.[42]

In the proposed integrated object model, the assembly relation supplies two kinds of information: the part name (ID) that the feature links, and the type of relationship between the feature and the connected part such as fit. Therefore, the related parts and feature links in the "place-transition" P/T model can be identified for every part. The system then uses a set to collect all the related parts. In the collection, the first position is filled with the part, and the remaining positions are filled with its related parts and links. Linking the collections of every part by features in a product forms the "place-transition" net graph of the product. In Fig. 6, the related parts p_3 and p_5 are identified through the feature, FixFit2; and the related part p_6 is found through feature keyFit. Similarly, all related parts are collected in a set $(p_1, p_2, p_3, p_4, p_5, p_6, p_7, p_8)$ for all the remaining features of the part as shown in Fig. 6. Based on the "place-transition" P/T net model, an assembly can be represented by the topological structure of its places (parts) and the liaison relationships (transitions and arcs) between parts.

(2) Geometric liaison model

Depending on the feature geometry, an assembly of any two connected components of a product may be defined with one of the many possible assembly relationships, some of which are enumerated as: pressure fit (PrFit), push fit (PuFit), moveable fit (MFit), loose fit (LFit), taper fit (TFit), spline fit (SPFit), ring fit (RFit), key fit (KeyFit), screw fit (ScrFit), contact fit (Contact), etc. The variety of assembly relations combined with an unlimited possible part geometries make it extremely difficult to develop a unique mathematical model to solve the assembly problem.

The product assembly relations above can be grouped into two classes: *fit* and *contact*, as shown in Figs. 8(a) and (b). With the help of feature information, these two classes of assembly relations between parts can be converted into a single representation contact relation. Through the assembly relation between a feature and a related part, contact relations in different directions can be defined. For example, in Fig. 8(a), part B fits into part A along the $+X$ direction (assume that the fitting axis is along the X axis) through its feature f_B; in Fig. 8(b), part B contacts with part A through its feature f'_B along the $+X$ direction, and part A contacts with part B through its feature f'_A along the $-X$ direction. Therefore there is a contact in any direction perpendicular to the fitting axis (X). Considering the directions parallel to the three coordinate axes, a fit relation can be replaced by

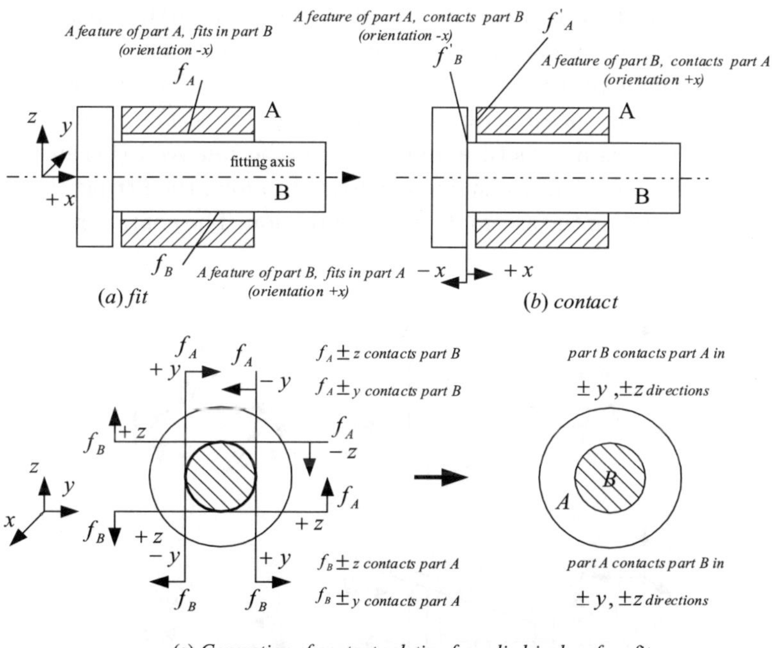

(c) *Generation of contact relation for cylindrical surface fits*

Contact	+x	-x	+y	-y	+z	-z
Part A		B	B	B	B	B
Part B	B		A	A	A	A

(d) *Summary of contact relation for cylindrical surface fits*

Fig. 8. Contact and fitting relations.[42]

contact relations in the six directions in a coordinate system $\pm X$, $\pm Y$, $\pm Z$, as shown in Fig. 8(c). The representation of the relationships between parts A and B in contact relations in the defined direction can be summarized as in Fig. 8(d).

Fitting relations such as pressure-fit, push-fit, position-fit, movable-fit, screw-fit, tape-fit, and ring-fit can be substituted with contact relations in the directions perpendicular and/or parallel to the fitting axis along the three coordinate axes. The first three fitting relations above (pressure-fit, push-fit, position-fit) have a common characteristic of contacting between two cylindrical surfaces. This allows for the examples shown in Fig. 8 to be used directly for generating contact relations.

For the ease of screw fit and tape fit, the contact relations do not confine to the directions perpendicular to the fitting axes, but also in other directions. The contact direction of a tape-fit as shown in Fig. 9 are N_A, N_B, N'_A, N'_B between parts A and B. These directions are made up by two of the six directions: $N_A = \{-Z, +X\}$, $N_B = \{+Z, -X\}$, $N'_A = \{-Y, +X\}$, $N'_B = \{+Y, -X\}$, where $+X$ and $-X$ are the directions along the fitting axis. The contact relationships between parts A and B exist in the directions perpendicular to the fitting axis and along the fitting axis. Figure 9 summarizes the representation of the contact relationships between parts A and B. Whether there is a contact relation along the fitting axis of a screw fit depends on the type of female parts. Except for the case of a screw nut, the contact relations of the female part may exist not only in the directions perpendicular to the fitting axis, but also in the direction along the fitting axis. For example, the female part A in Fig. 10(a) is a structure and it cannot be removed (because of its size, weight, etc.). It is equivalent to having a constraint along the fitting axis. A contact relation also exists along the assembly fitting direction. The contact relations defined

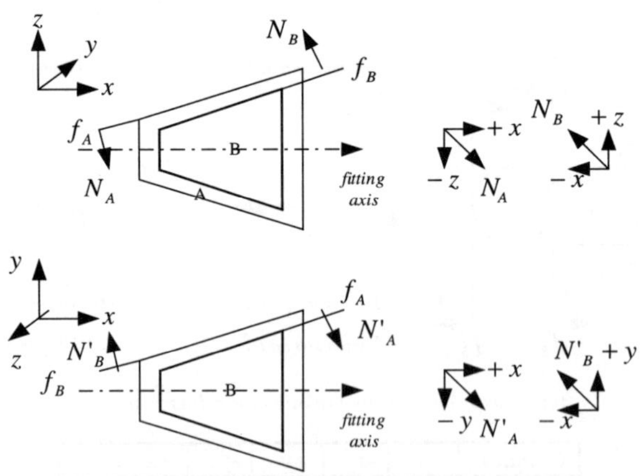

Contact	+X	-X	+Y	-Y	+Z	-Z
Part A	B			B		B
Part B		A	A		A	

Fig. 9. Generation of contact relations for tape fit.

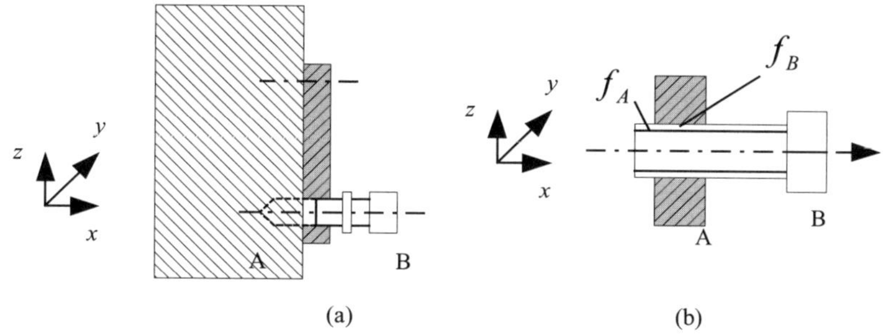

Fig. 10. Generation of contact relations for screw fit.[42]

here provide two types information: the connective relationship between parts and the removal constraints in the contact directions. Part A in Fig. 10(a) cannot be moved in the $-X$ direction when part B is fixed. This implies that part A has a moving constraints in the $-X$ direction. An additional contact relation needs to be added between parts A and B in the $-X$ direction along the fitting axis, i.e. part A contacts part B in the $-X$ direction. If the female part is simply a screw-nut as in Fig. 10(b), both the screw bolt and the nut can be removed to disassemble the pair, which is similar to the contact relationship of two cylindrical surfaces. For a unified representation in the system, a contact constraint along the fitting axis for the female part is defined. As such the screw-nut in Fig. 10(b) is defined to have a contact to the screw-bolt in the $-X$ direction. This extra contact relation between a screw-bolt and a screw-nut associated with the heuristic algorithms ensures that they will be disassembled as a pair in the generated disassembly sequence. The extra contact relation is defined based on the degree of difficulty in the disassembling of one part from the other.

The contact graphs represent both geometric and non-geometric constraints for the parts in a product. Using the above method, the contact relations of the parts in assembly (Fig. 6) can be generated, as shown in Table 1. A product is assumed to be suitable for robotic assembly if it has rigid parts interconnected with each other in mutually orthogonal directions. Each part can be assembled by a simple insertion or fastening such as screwing. The complete liaison model of an n-part assembly is a two-tuple $\langle \mathrm{P}, \mathrm{L} \rangle$,[25] where $\mathrm{P} = \{p_1, p_2, \ldots, p_n\}$ is an assembly and each symbol in P corresponds to one part in the assembly; and $\mathrm{L} = \{l_{ab} | a, b = 1, 2, \ldots, m, a \neq b\}$ is a set of 4-tuples, representing the relations between the parts in the assembly, where $(n-1) \leq m \leq n(n-1)/2$. The liaison l_{ab} represents the connective relations between a pair of parts p_a and p_b, where $p_a, p_b \in \mathrm{P}$. The connective relations can be divided into a contact-type and a fit-type connection. The assembly directions are defined with respect to the $\pm X$, $\pm Y$, $\pm Z$ directions. Then, a liaison l_{ab} can be expressed by a predicate as follows:

$$l_{ab} = liaison\langle p_a, C_{ab}, F_{ab}, p_b \rangle.$$

Table 1. Contact relations between parts of axle system.

Part	Contacts $+X$	Contacts $-X$	Contacts $+Z$	Contacts $+Y, -Y, -Z$
p_1	$p_2, p_4, p_7,$	p_3, p_8	$p_2, p_3, p_4, p_5, p_6, p_7$	$p_2, p_3, p_4, p_5, p_6, p_7$
p_2	p_4	p_1	p_1, p_{13}	p_1, p_{13}
p_3	p_1	p_5	p_1, p_{13}	p_1, p_{13}
p_4	p_7	p_1	p_1	p_1
p_5	p_3	p_6	p_1	p_1
p_6	p_5	p_8, p_9, p_{10}	p_1, p_9, p_{10}	p_1, p_9, p_{10}
p_7	p_{12}	p_4	p_1	p_1
p_8	p_6	p_{11}	p_9, p_{10}	p_9, p_{10}
p_9	p_8		p_8, p_6	p_8, p_6
p_{10}	p_8		p_8, p_6	p_8, p_6
p_{11}	p_8			
p_{12}		p_7, p_1	p_1	p_1
p_{13}	p_2	p_3	p_2, p_3	p_2, p_3
p_{14}	p_8	p_9	p_9	p_9

The C_{ab} is called a "contact-type" connection matrix, and the l_{ab} is a "fit-type" connection matrix. $C_{ab} = (c_{+x}, c_{+y}, c_{+z}, c_{-x}, c_{-y}, c_{-z})$ is a 1×6 binary function representing the contacts between the parts p_a and p_b, and $F_{ab} = (f_{+x}, f_{+y}, f_{+z}, f_{-x}, f_{-y}, f_{-z})$ is a 1×6 binary function representing the fit-type (translational motion) between part p_a and part p_b. C_{ab} and F_{ab} are known as contact functions, or C-function and a translational function, or F-function respectively. The element c_d of C_{ab} is represented by 0 for no contact, r_c for real contact, and v_c for virtual contact in the d direction between p_a and p_b where $d(+X, +Y, +Z, -X, -Y, -Z)$. Similarly the element f_d of F_{ab} is represented by 0 for no fit, s_w for screwing, r_f for round peg-in-hole, and m_p for multiple round peg-in-holes. For simplicity, C_{ab}: $c_d\{0, 1\}$, $c_d = 0$ indicates the absence of contact in that direction; $c_d = 1$ indicates that part p_b is in contact with part p_a in the direction of d. F_{ab}: $f_d \rightarrow \{0, 1\}$, $f_d = 1$ indicates that part p_b has the freedom of translational motion with respect to part p_a in the direction of d; $f_d = 0$ indicates the part p_b has no freedom of translational motion with respect to part p_a in the direction of d. From the C and F functions representing the relations such as contact and mobility among the parts, the assembly incident matrix of "place-transition" model can also be easily generated.

3.5.2. *Constraint models for assemblies*

An assembly procedure is usually subjected to the constraints of different parts and subassemblies. These constraints can be broadly classified into two groups: hard constraints and soft constraints. The hard constraints are the geometric and physical constraints related to the generation of assembly sequence, and the soft constraints are made by assembly planners related to the selection and evaluation of assembly sequence. These are discussed below.

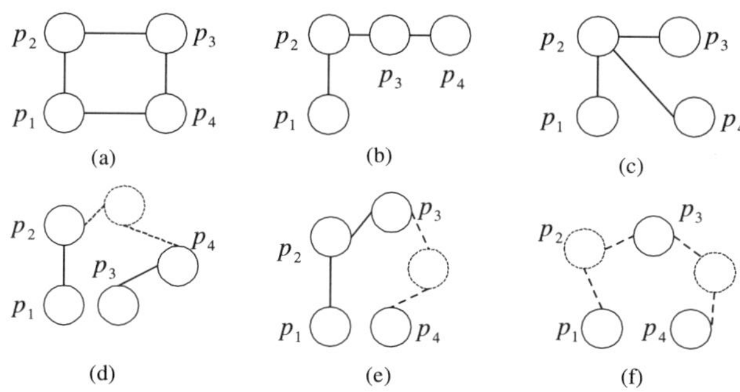

Fig. 11. Local liaison graph and its coherence.

(1) Topological constraints

Assembly topological constraints imply that two parts are interconnected or at least one part is connected with another part in the subassembly directly. The existence of topological constraint for a subassembly implies that the coherence of its local liaison graph is defined such that there is at least one path from an arbitrary part in a subassembly to any other parts in this subassembly.[91] For example, the local liaison graphs composed of p_1, p_2, p_3 and p_4 are shown in Fig. 11, in which (a), (b) and (c) are coherent, but (d), (e) and (f) are not coherent. Therefore, this type of constraint can be easily identified from the P/T net, and it can also be described by the liaison matrix or sub-matrix.[91] If the rank of matrix **R** is *rank*(**R**), then the coherence of liaison graph net can be calculated as follows:

$$rank(R) = \begin{cases} n-1 & \text{liaison graph is coherent} \\ < n-1 & \text{liaison graph is not coherent} \end{cases}$$

The connectives between a pair of parts in an assembly can be expressed as contact functions. The entries of C-function (1 or 0) can be extracted from the relational models as explained above. The predicates such as *contact*, *not_contact*, *connect*, *coherent*, *not_connect*, *not_coherent*, shown in Table 2, are added to simplify the description of the topological constraint knowledge.

(2) Geometric constraints

Geometric models of individual parts are referred to as representations of their geometric attributes including relative locations and orientations, vectors or axes of rotation and translation in the world coordinate space. Geometric constraints of an assembly entail whether there are relative or allowable positions and orientation relations between two parts or a part and a subassembly or two subassemblies, and the collision-free paths in the assembly. The complexity of solutions to assembly planning can be substantially reduced by considering geometric constraints. In general, geometric feasibility constraints and the accessibility constraints,[51] can be used in the generation of assembly sequence. The relational models defined in the product restrict the parts' degrees of freedom in the assembly. For example, a peg which

Table 2. Predicate representation of assembly constraints.

Constraints	Predicates	Explanation	Examples
Topological constraints	*connect*	Connections exist between two parts or a part and a subassembly or two subassemblies.	$connect(p_1, p_3)$ $connect([p_1, p_2], [p_3, p_4])$ $connect(p_1, [p_1, p_2, p_3])$
	coherent	An assembly or a subassembly is coherent.	$coherent([p_1, p_2, p_3])$
	not_connect	negative *connect*	$not\_connect\ (p_1, p_3)$
	not_coherent	negative *coherent*	$not\_coherent\ ([p_1, p_2, p_3])$
Geometric constraints	*interfer* (*geo_unfeasible, inaccessible*)	Interference exists between two parts or two subassemlies or a part and a subassembly.	$interfer(p_1, p_3)$ $interfer([p_1, p_2], [p_3, p_4])$ $interfer\ ([p_1, p_2], [p_3, p_4])$ $interfer(p_1, [p_1, p_2])$
	position	3D position coordinates	$position(p_i, x, y, z)$
	orientation	3D orientation coordinates	$orientation(p_i, \phi, \varphi, \psi)$
Partial precedence constraints	*precede* *e.g. base*	A part or a subassembly is assembled with other parts or subassemblies in a desired direction or precedence	$precede(p_1, p_2)$ $precede(p_1, [p_1, p_2])$ $precede([p_1, p_2], [p_3, p_4])$ etc.
Stability and security constraint	*unstable*	A subassembly without stability	$unstable(p_1, p_2, p_3)$
	changeable	A subassembly without security	$changeable(p_1, p_2, p_3)$
Cost constraints	*time_consuming*	An assembly operation takes much time than others	$time\_consuming(p_1, [p_2, p_3])$ $time\_consuming([p_1, p_2], [p_3, p_4])$
	cost_effective	The cost of a subassembly or a subassembly operation	$cost\_effective\ ([p_1, p_2, p_3])$ $cost\_effective\ (p_1, [p_2, p_3])$

loosely fits into a hole has only two degrees of freedom: translation along the hole axis and rotation about it. This information is modeled according to the parts in the assembly knowledge base. This can be extracted from the geometric data available in a geometric modeling system or from the inquiries directed to the user.

Based on an analysis of the translational degrees of freedom of the two parts involved in the defined relations, the geometric feasibility of separating two subassemblies in a disassembly operation can be computed automatically. Formally, if there does not exist any disassembly direction along an axis, this operation is not geometrically feasible. Furthermore, an operation is geometrically feasible if the intersection of all the disassembly directions of all incident parts in either one of the two subassemblies resulting from a subset is not empty.[51]

However, the restricted access of a tool to hold and remove a part or a subassembly may make the corresponding disassembly operation difficult to execute. In addition a geometrically feasible disassembly operation may result in the inaccessibility of the moved subassembly. It is therefore important to consider the accessibility constraints in assembly sequence planning. The formulation of this constraint is complex and difficult to compute automatically. A method for indirectly incorporating such constraints is described in partial precedence constraints below.

The degree of difficulty in disassembling a part depends on the number of parts that obstruct the path of the part to be removed. The more part is obstructed, the more difficult it is to move the part. Some of these may have contact relations with the part; while others do not (they are non-touch constraints for the part to be removed). Therefore, the strategy for searching a collision-free path in a given direction is divided into two stages.[42] The first stage is the multiple target elimination in the search for a part that does not contact any other parts in the given direction (contact-free part). The second stage is the non-touch constraints identification to check if the part found at the first stage has non-touch constraints. The part being searched for at the first stage is known as a target part and the target part that passes the second stage checking is known as a candidate part.

On the one hand, a target part which is contact-free in a direction can be found directly from contact relation graphs. It is possible that there are simultaneously several target parts available in one direction. In this case, based on the feature locations of the target parts,[42] the following heuristic algorithms are used to find the best target. To illustrate the algorithm, it is assumed that N is the direction in which the target part has a contact-free path, and $\overline{N}$ is the opposite direction of N.[42]

Step 1: Among the features of each target part, search for the feature that contacts another part in the $\overline{N}$ direction and is the farthest along the $\overline{N}$ direction.

In Fig. 12(a), part p_7 is one of the target parts in the $+X$ direction; and it contacts part p_4 through f_1^7, f_2^7, and f_2^4 in the $-X$ direction. Among these features, f_2^4 is the farthest in the $-X$ direction.

Step 2: Among the features selected in the first step, search for the farthest one along the N direction. If a feature is found, then the part that has the feature is the best target. In Fig. 12(b), parts p_9 and p_8 are the target parts in the $-X$ direction; f_2^8 and f_2^9 are the farthest features contacting part p_6 in the $+X$ direction; f_2^9 is farther than f_2^8. Therefore, part p_9 is removed prior to part p_8.

Step 3: If no feature is found in Step 2, it means that all the features found in Step 1 are on the same plane. Therefore, among all the other features of those target parts, search for the feature that is farthest along the N direction. If the feature is found, then the part having this feature is the best target.

In Fig. 12(c), parts p_9, p_{10} and p_{11} (a virtual part) are the target parts in the $-X$ direction; f_2^9, f_2^{10} and f_2^{11} are on the same plane; f_3^{11} is the farthest features. Therefore, part p_{11} is removed prior to part p_9 and p_{10}.

Step 4: If no feature is found in Step 3, all the targets have the same priority, and any one of them can be removed first. Therefore, the best target is selected arbitrarily.

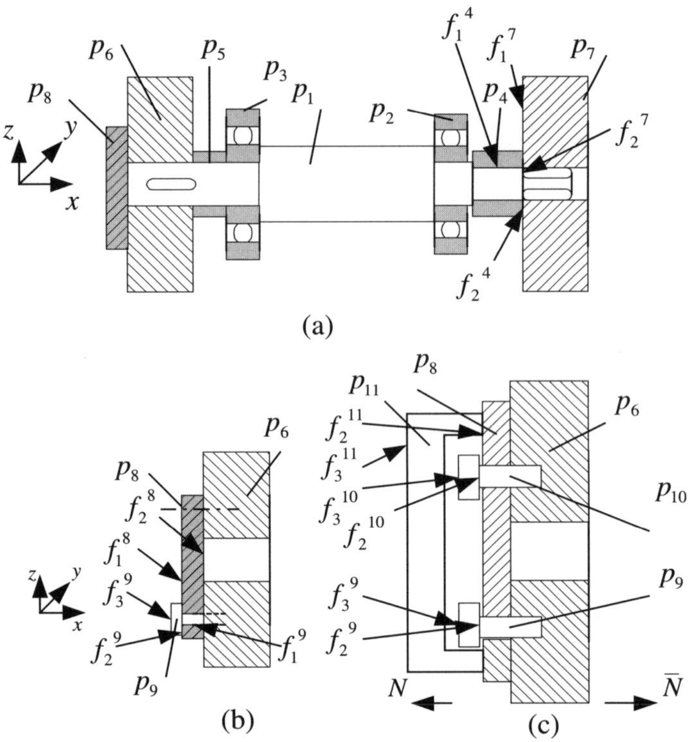

Fig. 12. Example geometric constraint analysis.

In Fig. 12(c), p_9 and p_{11} are all candidates; f_2^9 and f_2^{10} are on the same plane; f_3^9 and f_3^{10} are on the same position along the N direction. Therefore, p_9 and p_{10} will have the same priority.

On the other hand, when a part is contact-free in a direction, it does not imply that the part has a collision-free path in that direction. There may be a non-touch obstruction on the path of the part's removal. Therefore, for the best target just determined with the above algorithm, the following heuristic algorithm will be used to check if it has non-touch constraints in the contact-free direction N ($\overline{N}$ is opposite direction of N)[42]:

Step 1: Search for the feature F which is the farthest in N with a normal in $\overline{N}$ from the features of the parts that the target part does not contact in $\overline{N}$. If the feature is found to belong to the target, then the target part has a collision-free path in N; otherwise, if the feature found does not belong to the target part, then go to Step 2.

In Fig. 13(a), part p_{12} is the target in the $+X$ direction. Feature f_1^{12} of part p_{12}, whose normal in the $-X$ direction, is the farthest in the $+X$ direction. p_{12} is therefore a candidate part that has a collision-free path in the $+X$ direction.

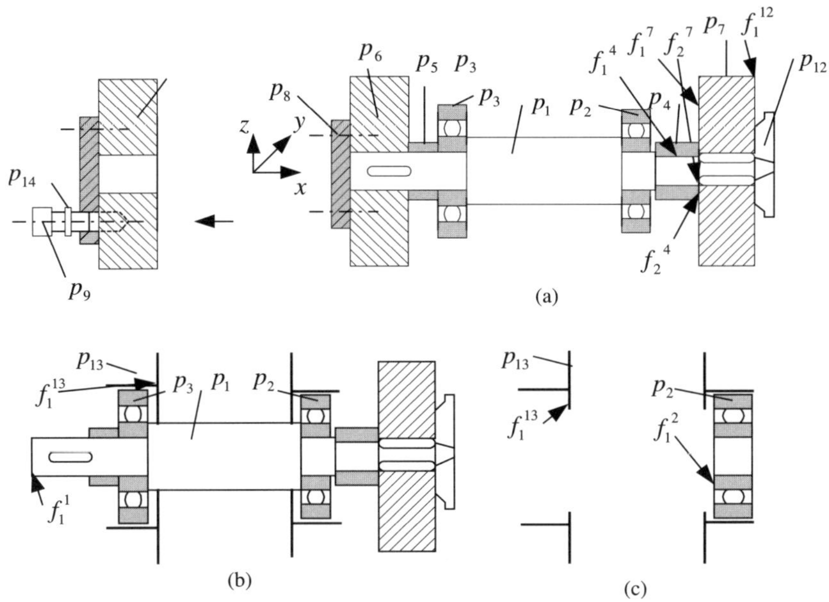

Fig. 13. Examples of qualified candidates with stability.

Step 2: Search for the feature f of the target which has its normal in N and is the farthest in N. In the N direction, if f is located farther than or equal to F, then the target part is selected; otherwise if F is located farther than f, then the target part is rejected. The system searches for another target part from other directions.

In Fig. 13(b), part p_1 is the target part in the $-X$ direction; f_1^{13} of part p_{13} and f_1^1 of part p_1 are the features found in Steps 1 and 2, respectively. Since f_1^1 is farther than f_1^{13}, part p_1 is selected. Similarly, in Fig. 13(c), part p_2 is the target part in the $-X$ direction; f_1^{13} of part p_{13} and f_1^2 are the features found in Steps 1 and 2, respectively. Since f_1^{13} is farther than f_1^2, part p_2 is rejected in the $-X$ direction.

After a part has been identified as a candidate, the determination of its removal is made by considering the stability of the assembly structure. This will be discussed in stability and security constraints below.

As described above, the qualitative representation of possible connectives, motions, and how they are constrained in three-dimensional space can be expressed by two binary functions, C-function and F-function, in the liaison model where the six entries represent the six directions of motion as described above. The values of C-function (1 or 0) can be extracted from the geometry and topology of the parts as explained above. A part p_a can be dismantled from another part p_b in the direction d, if and only if there exists a collision-free path in that direction. The existence or absence of a collision-free path for the disassembly of one part from the other can be conveniently expressed by the F-functions. The six entries of F_{ab} represent the six linear half (three translational) degrees of freedom along the principal axes of motion. The mechanism employed to extract the F-function is exactly the same as the C-function, except in this case where the parts are moved over a certain specified distance. If there is no interference in the disassembly path, the entry of F-function is 1, otherwise it is 0. Alternatively, geometric constraints are represented by predicates such as *position*, *orientation*, *interfer* (*geometry_unfeasible*, *inaccessible*), as shown in Table 2.

(3) Stability and security constraints

A subassembly is said to be stable if its parts maintain their relative positions and do not break contact spontaneously. This may be expressed in terms of gravity stability and anti-disturbance stability under the action of gravity, assembly stability under the action of assembly force, and plastic stability under the action of inner spring force. A secured subassembly has zero degree of freedom for the relative motion of all parts in the subassembly. A subassembly without stability or security constraints means that it is unstable and changeable, and thus unfeasible.

A subassembly has stability if it satisfies one of the following:

(1) parts in subassembly with constraints of fasteners;
(2) parts in subassembly with tight or clearance mating;
(3) parts in subassembly with its gravity center on the supporting surface of constraint; and
(4) each part in subassembly has stability.

A subassembly has security if it satisfies one of the following:

(1) the subassembly is fastened;
(2) parts in a subassembly has direct liaison with fasteners and has zero degree of freedom of relative motion; and
(3) parts in a subassembly has direct liaison with fasteners and has motion degree of freedom of fastening constraints only.

When the collision-free path of a part (the candidate determined above) is along the $+Z$ direction (which usually means that the candidate part does not support any other parts in the $+Z$ direction, except for a screw-bolt which fits with a screw-nut), the removal of the candidate will not affect the stability of an assembly structure. Therefore the candidate can be removed. In the $+Z$ direction, if the candidate part is a screw-bolt, then a screw-nut is searched from the related parts of the screw-bolt. If the search fails, then the screw-bolt is treated as other types of parts having collision-free paths in the $+Z$ direction. The procedures discussed above will apply. If the search is successful, then a temporary support is added to the screw-nut, and the screw-bolt is removed in the $+Z$ direction. The screw-nut is removed right after the screw-bolt is disassembled in the $-Z$ direction.

It is not common for a part to be removed in the $-Z$ direction because the $-Z$ direction is set the lowest priority for considerations.[42] It is recommended to set up a product in such a manner so that most of the parts can be removed without considering the $-Z$ direction. If a part (the candidate part) has a collision-free path only in the $-Z$ direction, this part can be removed without the stability examination except when the candidate is a screw-bolt. The reason is that if a candidate part other than a screw-bolt does not contact with any other parts in the $-Z$ direction, the candidate part does not have any support from the $-Z$ direction and is not a support part for the structure. Therefore, the candidate part can be removed. Otherwise, the structure itself may be unstable. When the candidate part is a screw-bolt, its function is usually to fasten another part onto some structures, and thus the removal of the screw-bolt will cause the part it contacts in the $+Z$ direction to be unstable. When this happens a temporary support is added to the unstable part, and then the screw-bolt is removed in the $-Z$ direction. The unstable part is removed right after the screw-bolt in the same direction.

When a collision-free path of a part (the candidate) is in a horizontal direction (e.g. $+X$, $-X$, $+Y$ or $-Y$), instability may occur if the candidate is removed. Therefore, when a part has a collision-free path in a horizontal direction, the decision on whether it can be removed will be made through a three-step heuristic check[42]:

Step 1: Check whether the candidate part contacts with any other parts in the $+Z$ direction. If the candidate part does not contact with any parts in the $+Z$ direction, it can be removed; otherwise, if the candidate part does contact some parts known at the virtual unstable parts in the $+Z$ direction, go to Step 2;

Step 2: For each of the virtual unstable parts, search for its contacting parts, also known as its supporting parts, in the $-Z$ direction. If the number of supporting parts is more than one, it means that the virtual unstable part has other support in addition to the candidate's support, then the candidate part can be removed. If the virtual unstable part has only one support, then the candidate is not approved and is then suspended in the candidate position known as a suspended candidate. The removal of the candidate part may cause other parts to become unstable and those virtual unstable parts to become real unstable parts. The system then searches for another candidate. If another one is found, the suspended candidate is taken out from the candidate position the system proceeds with the new one. If no more such candidate is found, the suspended candidate proceeds to Step 3.

Step 3: Add a temporary support to the unstable parts. Then the candidate part can be removed. The unstable part is removed right after the candidate. The removal direction of the unstable part is either the same as that of the candidate or is in the $-Z$ direction. If the former is tenable, then the unstable part and the candidate part are grouped into a sub-assembly in the final assembly sequence.

In the previous example shown in Fig. 13(a), part p_9 has a collision-free path in the $-X$ direction and is therefore a candidate part. Part p_{14} gets its only support from p_9. It is an unstable part. Therefore part p_9 is suspended, and the system searches for another candidate. In the $+X$ direction, part p_{12} is found as a candidate to be removed since p_7 contacts in the $-X$ direction also have contacts with other parts p_4 and p_1 in the $+X$ direction. In the meantime, part p_{14} is no longer a candidate, and the system proceeds from p_4 and p_1, then p_2.

In Fig. 13(b), part p_1 is a suspended candidate in the $-X$ direction since part p_5 has its only support from p_1. Therefore, the system searches for other possible candidates in other directions but failed. The unstable part p_5 gets a temporary support, and part p_1 is restored as an approved candidate and then removed in the $-X$ direction. Thereafter p_5 is removed in the $-X$ direction.

Similarly in Fig. 13(a), part p_9 is a suspended candidate in the $-X$ direction and part p_{14} is a unstable part. Since other candidate cannot be found, p_{14} gets a temporary support, and p_9 becomes an approved candidate and is removed in the $-X$ direction. p_{14} and p_9 can be grouped into a subassembly, that is, in the final assembly sequence, they can be treated as an unit to be assembled/disassembled to/from the assembly.

Based on the above discussions, the stability and the security constraints can be reasoned out indirectly from examination of the removal of candidate part along the collision-free path. In practice, they can also be interactively extracted from the information provided by the user. For this purpose, predicates such as *unstable* and *changeable* in Table 2 are used to represent them.

(4) Partial precedence constraints

Partial precedence constraints specify that a part or a subassembly has to be assembled with a part or a subassembly in a desired direction or precedence in terms of an assembly operation. This type of constraint is crucial for assembly planning. For example, it may be to place the base part of an assembly first and the expensive parts last, or it may be to place the heavy parts as soon as possible. It can easily be generalized that some parts should be placed at the beginning or at the end for reasons other than their cost or their weight. Therefore, the partial precedence constraints are the keys to the construction of a directed P/T net graph, the determination of the feasibility of subassembly and subassembly operations, and the generation of part or subassembly based sequences. Generally, they can be extracted from geometric and non-geometric information provided by the user interactively and are available from solid model in the CAD database.

The direction of an arc connecting two nodes p_a and p_b in a P/T net graph can be used to represent the partial precedence of assembly. This new representation scheme can be illustrated with the axle system as shown in Fig. 14. The following rules may be helpful in determining the direction and constructing the directed P/T net graph:

- If p_a is a connecting part, the direction is from p_b to other adjacent nodes;
- If p_a is inside p_b, or p_a is inserted into p_b, the direction is from p_a to p_b;
- If p_b is beneath p_a, the direction is from p_a to p_b;
- If p_b is a datum, p_a is connected with p_b on the side, the direction is from p_a to p_b.

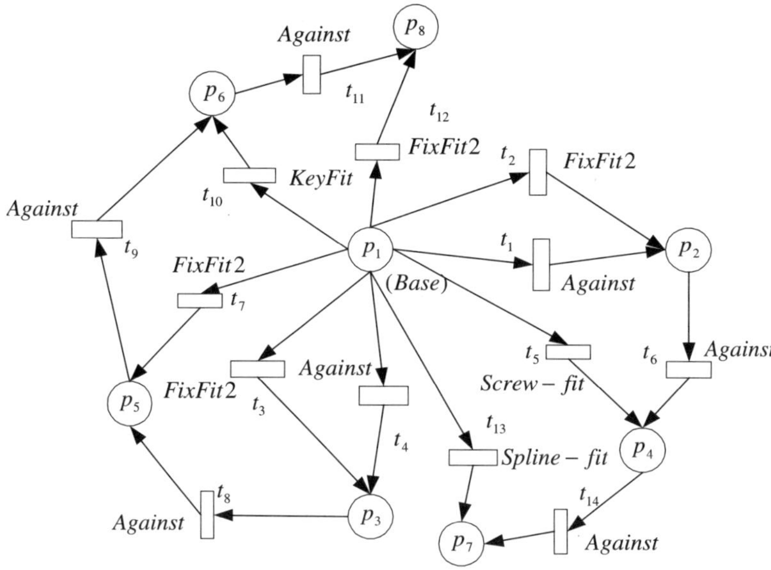

Fig. 14. The directed P/T net model of axle system.

Geometric reasoning can extract the precedence restrictions imposed by the hard constraints. There are two steps to obtain the relationships among the parts of an assembly: checking the existence of contact between each pair of parts; and checking for possible disassembly directions of each part. For each pair of mating parts, the feasible mating directions can be determined by considering geometric constraints from the mating faces between these two parts. Collision constraints may exist for each pair of non-mating parts. This collision information can be obtained by performing a collision detection between solid models of the parts. The collision information for an assembly needs to be determined for both mating and non-mating parts.

To reason about the partial precedence, the following motion constraints need to be considered: translation and rotation constraints. For each type of motion, there are three directions and two senses, ($\pm X$, $\pm Y$, $\pm Z$), hence there are six motions that a part can exhibit. The notion of direction and sense must be employed to reason about the spatial relationships existing among the parts of an assembly. The proposed approach operates with parts positioned in a tri-orthogonal Cartesian coordinate system. Every part is oriented along the three principal axes. The priority order of the six directions is determined by considering the stability of the structure in the disassembly process and the ease of automated assembly operations. Since the removal of a part from top to bottom is known as the most preferable disassembly operation, the $+Z$ assembly direction is considered to be the most stable orientation of the product assembly, having the highest priority. The direction $-Z$ has the lowest priority, since it usually requires either the product to be turned over or a special fixture to be used. The priority order among $\pm X$ and $\pm Y$, which are horizontal directions, cannot be determined at this early stage based on the above considerations. The strategy temporarily set the priority order among them: $+X \rightarrow -X \rightarrow +Y \rightarrow -Y$. After the sequence is generated, the order of these four directions is then reanalyzed and determined according to the number of parts disassembled in a direction based on the ease of the automated assembly operation. The bigger the number of the parts disassembled in a direction is, the higher the priority of the direction is. The priority order of the six direction is therefore $+Z \rightarrow +X \rightarrow -X \rightarrow +Y \rightarrow -Y \rightarrow -Z$. As a result, the precedence relations among parts are first determined by the priority of the direction that a part can be removed. For example, if part A can be removed in the $+Z$ direction and another part B can be removed in the $+X$ direction, then part A is removed prior to part B; if part C can be removed either in the $+Z$ direction or in the $+X$ direction, then part C is removed in the $+Z$ direction.

A precedence constraint of a liaison l_{ab} is represented by a set of nP parts that must be connected before two parts p_a and p_b are interconnected. The precedence constraint of a liaison l_{ab} is expressed by a predicate *precede*(l_{ab}) as follows:

$$\textit{precede}(l_{ab}) = \{p_c | c = c_1, c_2, \ldots, c_{nP}\}.$$

As a dual representation of the above equation, a set of nL liaisons that have a part p_c as an element of *precede*(l_{ab}) can be defined. This set of liaison $LS(p_c)$ is expressed by:

$$LS(p_c) = \{l_{ab} | a_1, a_2, a_{nL}, b = b_1, b_2, \ldots, b_{nL}\}.$$

Alternatively, a set of nQ parts having precedence constraints with a part p_a can also be defined. This set of parts $PS(p_a)$ is expressed by:

$$PS(p_a) = \{p_b | b = b_1, b_2, \ldots, b_{nQ}\}.$$

An example of the precedence constraints for the subassembly in Fig. 12 are inferred as:

$$\textit{precede}(l_{15}) = \{p_3\},\ \textit{precede}(l_{16}) = \{p_5, p_3\},\ \textit{precede}(l_{27}) = \{p_4\},\ \textit{precede}(l_{41}) = \{p_2\}$$

Also, the set of liaisons and the set of parts are derived as:

$$LS(p_2) = \{l_{41}\},\ LS(p_3) = \{l_{15}, l_{16}\}, LS(p_4) = \{l_{27}\},$$
$$LS(p_5) = \{l_{16}\},\ PS(p_5) = \{p_1, p_3\},\ PS(p_7) = \{p_2, p_4\}.$$

Besides the C and F functions above, the predicates such as *precede* in Table 2 is further used to represent some partial precedence relations in an assembly so as to interactively simplify and help the geometric reasoning. The partial precedence relations has properties as follows:

(i) Chain property: $\textit{precede}(p_1, p_2) \vee \textit{precede}(p_2, p_3) \vee \textit{precede}(p_3, p_4) \Rightarrow \textit{precede}(p_1, p_4)$;
(ii) Commutative property: $\textit{precede}(p_1, p_2) \equiv \textit{precede}(\bar{p_2}, \bar{p_1})$;

(iii) Distributive properties:

$$precede(p_1 \wedge p_2, p_3) = precede(p_1, p_3) \wedge precede(p_2, p_3);$$
$$precede(p_3 \wedge p_1, p_2) = precede(p_3, p_1) \wedge precede(p_3, p_2);$$
$$precede(p_1 \vee p_2, p_3) = precede(p_1, p_3) \vee precede(p_2, p_3);$$
$$precede(p_3 \vee p_1, p_2) = precede(p_3, p_1) \vee precede(p_3, p_2),$$

where, "$\wedge$" represents *AND*; "$\vee$" represents *OR*; and "$-$" represents *NOT*. The precedence constraint is the key to the determination of the feasibility of subassembly and the subassembly operation and the generation of part-based and subassembly-based sequence.

(5) Deductive rules for constraints reasoning

To simplify the reasoning process of constraints in assembly planning, the following basic rules can be deduced and should be applied in the procedure of automatic reasoning, where "$\Leftrightarrow$" means "*equivalent*" and "$\Rightarrow$" means "*implies*".

(i) *Equivalent rule:* a subassembly is a compound of at least one part. Thus, it is not associated with the location of parts in the subassembly. For example, $[p_1, p_2, p_3, p_4] \Leftrightarrow [p_1, p_2, p_4, p_3] \Leftrightarrow [p_1, p_4, p_3, p_2] \Leftrightarrow \cdots$ and $[[p_1, p_2], [p_3, p_4, p_5]] \Leftrightarrow [[p_1, p_2], [p_3, p_5, p_4]] \Leftrightarrow [[p_2, p_1], [p_4, p_5, p_3]] \Leftrightarrow \cdots$.
(ii) *Topological constraint existence rule*: the local liaison graph for a subassembly with topological constraints must be coherent.
(iii) *The Loop-closure rule*: a cycle in a liaison graph implies a need to simultaneously complete two of the liaisons in the cycle. It results from assuming rigid parts and liaisons.
(iv) *Subassembly operation rule*: if a subassembly is unfeasible then its all decomposed assembly operations are not available.
(v) *Superset rule*: if there is no mating between two parts or subassemblies or one part and one subassembly due to interference in the approaching path, then, the adding of a part or a subassembly to either a part or a subassembly in the original assembly, which is not associated with the mating liaisons, will not change this case. For example, $not\_connect(p_1, p_2) \wedge interfer([p_1], [p_2]) \wedge not\_connect(p_3, p_2) \Rightarrow not\_connect([p_1, p_3], [p_2])$.
(vi) *Subset rule*: if a mating can occur in two subassemblies, then removing a part from either subassembly, which is not associated with the mating liaison(s), will not change this case. For example, $connect([p_1, p_2], [p_3, p_4]) \wedge connect(p_2, p_3) \wedge not\_connect(p_1, p_3) \Rightarrow connect([p_2], [p_3, p_4])$.

4. Assembly Sequence Generation and Visualization

Similar to the combinatorial optimization problems such as time table, graph coloring, job shop scheduling, and vehicle routing, assembly sequence planning can be in nature described as a constraint satisfaction problem (CSP). It is NP-complete and at huge instances requires a large amount of computational resources. This is the reason why it is very important in the field of AI, with much research centered on the discovery of new algorithms. The proposed CSP framework and algorithm for assembly sequence planning can offer a sound basis for representing and solving decision problems with/without uncertainty.[89]

4.1. *Determination of feasible subassemblies*

From the assumption above, the whole assembly procedures of an n-part product can be classified into $n-1$ levels. Each subassembly consists of a primitive part. Level 0 is a primary subassembly and level $n-1$ is the highest level which forms the final product. Level $i-1(i=2,\ldots,n-1)$ is the subassembly that consists of at least two up to $n-1$ parts.

4.1.1. *All subassembly configurations*

The number of theoretical subassembly configurations in level $i-1$ is C_n^i, where $i=1,2,\ldots,n$, which is determined by compound principles in mathematics. The theoretical configuration numbers and all configurations are shown in Tables 3 and 4.

4.1.2. *Unfeasible subassemblies*

In spite of the large number of theoretical sequences, only a few of them are feasible. The key to deciding on the unfeasibility of a subassembly is to test and check the constraint conditions. First of all, the topological constraint must be tested by checking the coherence of the *local liaison graph* and its liaison sub-matrix of a subassembly or the existence of the predicates including *connect*, *coherent*, *not_connect*, and *not_coherent*. Thereafter other constraints are checked. This is carried out until all constraints are tested by checking the corresponding predicates in assembly knowledge base for all subassemblies. It can be accomplished by the way of recursive decomposing and backtracking in artificial intelligence (AI). Finally all unfeasible subassemblies can be identified and removed.

Table 3. Theoretical number of subassemblies.

Number of parts	Assembly levels	Total number of subassemblies
1	0	1
2	0,1	3
3	0,1,2	7
4	0,1,2,3	15
5	0,1,2,3,4	31
6	0,1,2,3,4,5	63
7	0,1,2,3,4,5,6	127
8	0,1,2,3,4,5,6,7	255
...	...	...

Table 4. Theoretical subassembly configurations of n-part product.

Level	Theoretical subassembly configurations
0	$[p_1], [p_2], [p_3], \ldots, [p_n]$
1	$[p_1, p_2], [p_1, p_3], [p_1, p_4], \ldots, [p_1, p_n]$ $[p_2, p_3], [p_2, p_4], \ldots, [p_2, p_n]$ $[p_3, p_4], [p_3, p_5], \ldots, [p_3, p_n]$ ...... $[p_{n-2}, p_{n-1}], [p_{n-1}, p_n]$ $[p_{n-1}, p_n]$
......	......
$i-1$ $(i = 3, 4, \ldots, n-2)$	$[p_1, p_2, \ldots, p_{i-1}, p_i], [p_1, p_2, \ldots, p_{i-1}, p_{i+1}], \ldots, [p_1, p_2, \ldots, p_{i-1}, p_n]$ $[p_1, p_2, \ldots, p_i, p_{i+1}], [p_1, p_2, \ldots, p_i, p_{i+2}], \ldots, [p_1, p_2, \ldots, p_i, p_n]$ ......
$n-1$	$[p_1, p_2, p_3, \ldots, p_n]$

To realize the automatic reasoning, the predicates such as *test_constraint*, *exist* and *not_exist* are proposed to check whether the subassemblies and the subassembly operations satisfy all the above mentioned constraints and then determine if they are feasible or unfeasible.

4.1.3. *Recognition of feasible subassemblies*

Based on constraint knowledge, all unfeasible subassemblies of level $i-1$ can be eliminated, and the feasible subassembly subsets of this level can be formed. The number of feasible subassemblies at level $i-1$ can be obtained by the following formula: $N^i_{fs} = C^i_n - N^i_{infs}$, $i = 1, 2, 3, \ldots, n$, where N^i_{fs} is the number of feasible subassemblies at level $i-1$; C^i_n is the number of theoretical assembly configurations in level $i-1$; and N^i_{infs} is the number of unfeasible subassemblies at level $i-1$. The set of feasible subassemblies at level $i-1$ can be determined by the following formula: $\{S^i_{fs}\} = \{S^i_{ts}\} - \{S^i_{infs}\}$, $i = 1, 2, \ldots, n$, where $\{S^i_{fs}\}$ is a set of feasible subassemblies at level $i-1$, $\{S^i_{ts}\}$ is a set of theoretical subassembly configurations at level $i-1$, and $\{S^i_{infs}\}$ is a set of total unfeasible subassemblies at level $i-1$.

4.1.4. *Decomposition of feasible subassemblies*

A feasible subassembly in level $i-1$ is likely to be decomposed into many corresponding combinations of feasible subassemblies in lower level j ($j = 0, 1, 2, \ldots, i-2$), which are defined as assembly operations here. It is noted that subassemblies in lower levels should be feasible, otherwise this operation is unfeasible and should be removed.

4.2. *Assembly sequence searching and visualization*

Once the feasible subassemblies and their decompositions are available, the assembly operations can be determined, and the assembly sequence can be represented and visualized. In this section, a formal model based on the high-level Petri nets is used to represent and visualize assembly processes and assembly sequences. Motivations for using the Petri nets stem from their clear and well-defined semantics as well as from the formal analysis tools and techniques available. Petri nets have an easy to understand graphical representation of the assembly processes and assembly sequences. This representation allows difficult concepts in assembly processes such as concurrency and synchronization to be visualised and simulated in an intuitive manner.

4.2.1. *Building assembly Petri net*

An assembly Petri net can be built directly from a lower level enumeration of precedence constraints or from an intermediate AND/OR representation generated by feasible subassemblies, assembly operations and operation paths of disassembly or assembly.[91] The latter approach has been studied in Refs. 77 & 78. An algorithm for mapping an AND/OR graph into a Petri net has been given in Refs. 69 & 77. AND/OR graphs can only express free-choice concurrency, thus they represent a less general model than Petri nets. Petri net can express both decisions involving conflict and non-determinism and concurrency in operation. Furthermore, it expresses a clean concurrent operation devoid of confusion.[80] To generate the assembly Petri net from the feasible subassemblies and assembly operations, the following rules should be used:[91]

(1) Each place in the Petri net graph corresponds to the parts and the feasible subassemblies;
(2) Each transition represents the feasible assembly operations; and
(3) The directed arcs, which link places and transitions indicate the relationships between subassemblies and assembly operations.

The assembly Petri net model can be formulated when all feasible assemblies and their decompositions are generated. It can be described by a set as follows:[91]

$$\mathrm{APN} = (\mathrm{P}_{fs}, \mathrm{T}_{fsop}, \mathrm{F}, \mathrm{W}, \mathrm{M}_0),$$

where $\mathrm{P}_{fs} = (p_{fs,1}, p_{fs,2}, \ldots, p_{fs,n})$ is a set of feasible subassemblies; $\mathrm{T}_{fsop} = (t_{fsop,1}, t_{fs,2}, \ldots, t_{fsop,n})$ is a set of feasible assembly operations, $\mathrm{P}_{fs} \cap \mathrm{T}_{fsop} = \phi$, $\mathrm{P}_{fs} \cup \mathrm{T}_{fsop} \neq \phi$; $\mathrm{F} \subseteq (\mathrm{P}_{fs} \times \mathrm{T}_{fsop}) \cup (\mathrm{T}_{fsop} \times \mathrm{P}_{fs})$ is a set of arcs (flow relation); $\mathrm{W}\colon \mathrm{F} \rightarrow \{w_1, w_2, \ldots\}$ is an association weight function on arcs, $\forall\, f \in \mathrm{F}$, $\mathrm{W}(f) = w_i$, w_i is the weight of arc f. $\mathrm{M}_0\colon \mathrm{P} \rightarrow \{0, 1, 2, \ldots\}$ is the initial marking.

Figure 15(a) shows the parts to be assembled into a flashlight. Figure 15(b) shows the corresponding AND/OR graph expressing precedence constraints.[12,73] From the assumption, the assembly and disassembly sequences are the reverse each other. Therefore, it is only needed to construct either a disassembly or assembly

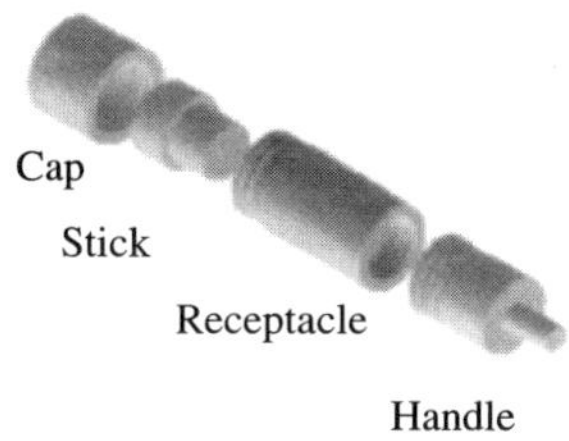

(a) The flashlight components

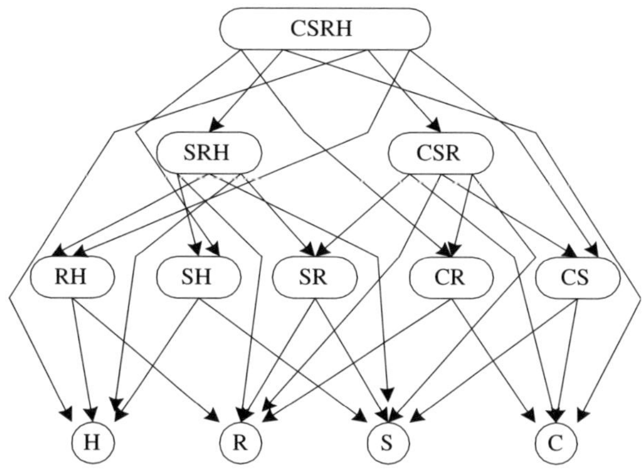

(b) AND / OR graph

Fig. 15. Flashlight and its assembly AND/OR graph.

Petri net. A simplified algorithm can be described as follows[69,76]:

Step 1: Convert each AND/OR node i into a place p_i; let p_{product} be the place corresponding to the assembled product and p_{ik}, $k = 1, \cdots, K$, is the place corresponding to the R individual parts, i.e. the leaves in the AND/OR graph;

Step 2: Convert each OR branch from a node i to other nodes l and from m into an (outgoing) transition t_j such that $^\bullet t_j = (p_i)$ and $t_j^\bullet = (p_l, p_m)$.

Step 3: Add a transition t_{loop} such that $^\bullet t_{\text{loop}} = (p_{pk} | k = 1, \cdots, K)$ and $t_{\text{loop}}^\bullet = (p_{\text{product}})$.

Step 4: Assign an initial marking M_0 such that $\mathrm{M}_0(p_{\text{product}}) = 1$ and $\mathrm{M}_0(p) = 0$, $\forall\, p \in \mathrm{P}$, $p \neq p_{\text{product}}$.

Transition t_{loop} serves the purpose of permitting a repetitive behavior and thus simplifies the steady state analysis. The initial marking M_0 corresponds to the beginning of the disassembly operation. Note that in assembly Petri nets there are no self-loops, i.e. the net is pure. Furthermore, all arcs weights are either zero or one, and there are no inhibitor arcs, thus the net is ordinary. Figure 16 shows the

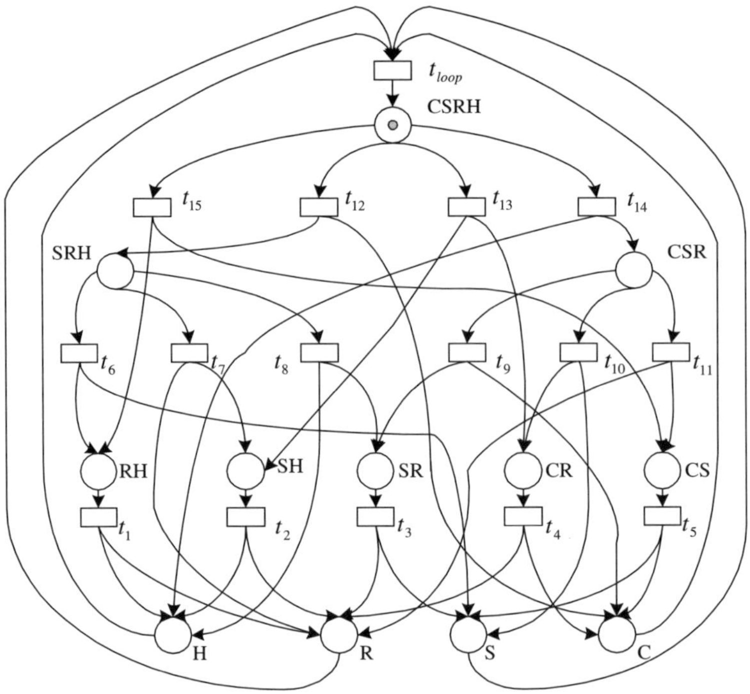

Fig. 16. The disassembly Petri net for the flashlight.

disassembly Petri net for the flashlight from the AND/OR graph in Fig. 15(b), where, $p_{\text{product}} = p_{12}$, $p_1 = \text{H}$, $p_2 = \text{R}$, $p_3 = \text{S}$, $p_4 = \text{C}$, $p_5 = \text{RH}$, $p_6 = \text{SH}$, $p_7 = \text{SR}$, $p_8 = \text{CR}$, $p_9 = \text{CS}$, $p_{10} = \text{SRH}$, $p_{11} = \text{CSR}$, and $p_{12} = \text{CSRH}$.

4.2.2. *Basic properties of assembly Petri nets*

More formally, Petri net representations of disassembly tasks are free-choice Petri net defined as an ordinary Petri net such that every arc from a place is either a unique outgoing arc or a unique incoming arc to a transition.[80] It is easy to see that this property holds for all Petri nets generated from AND/OR graphs by the algorithm above. The basic properties of ordinary Petri nets and free-choice Petri nets, such as boundedness, liveness, and safeness, are described in Refs. 76 & 80. A disassembly Petri net is an free-choice net. Its deadlocks and traps are given by the P-semiflows, which are properly marked in M_0. By applying the property analysis of ordinary Petri nets and free-choice Petri nets to assembly Petri nets, the following results are available for the assembly Petri nets[69,76]:

(1) Boundedness

The P-semiflows of a disassembly Petri net developed according to the algorithm above correspond to the potential subassemblies including a part. There is exactly one P-semiflow for each part, and then the net is covered by P-semiflows. Thus the disassembly Petri net is structurally bounded. As P-semiflows of disassembly Petri nets are boolean vectors, they can be denoted by the sets of places associated to non-null elements. The P-semiflows for the flashlight disassembly Petri net are the following[76]:

$$PS_1 = \{p_3, p_6, p_7, p_9, p_{11}, p_{10}, p_{12}\}, \quad PS_2 = \{p_2, p_5, p_7, p_9, p_{11}, p_{10}, p_{12}\},$$
$$PS_3 = \{p_1, p_5, p_6, p_{10}, p_{12}\}, \quad PS_4 = \{p_4, p_8, p_9, p_{11}, p_{12}\}.$$

(2) Safeness

The initial marking M_0 is such that $M_0(p_{\text{product}}) = 1$ and $M_0(p) = 0$, $\forall p \in P$, $p \neq p_{\text{product}}$. p_{product} belongs to all P-semiflows. There is exactly one token in any P-semiflow of the disassembly Petri net. Thus the disassembly Petri net is safe. Each P-semiflow of the flashlight disassembly Petri net yields an invariant relationship[76]:

$$M(p_3) + M(p_6) + M(p_7) + M(p_9) + M(p_{11}) + M(p_{10}) + M(p_{12}) = 1.$$

This relationship states that at any time the stick can only belong to one specific subassembly. Similar relationships hold for cap, receptacle, and handle.

(3) Liveness

The T-semiflows of a disassembly Petri net correspond to the possible disassembly sequences. Each disassembly operation is included in at least one disassembly sequence, thus the necessary condition for liveness is satisfied, i.e. a disassembly Petri net is live. Note that each disassembly operation only occurs once in a disassembly sequence. The T-semiflows for the flashlight assembly Petri net are as following[76]:

$$TS_1 = \{t_{12}, t_8, t_3, t_{\text{loop}}\}, \quad TS_2 = \{t_{12}, t_7, t_2, t_{\text{loop}}\}$$
$$TS_3 = \{t_{12}, t_6, t_1, t_{\text{loop}}\}, \quad TS_4 = \{t_{14}, t_{11}, t_5, t_{\text{loop}}\}$$
$$TS_5 = \{t_{15}, t_1, t_5, t_{\text{loop}}\}, \quad TS_6 = \{t_{14}, t_{10}, t_4, t_{\text{loop}}\}$$
$$TS_7 = \{t_{14}, t_9, t_3, t_{\text{loop}}\}, \quad TS_8 = \{t_{13}, t_2, t_4, t_{\text{loop}}\}$$

(4) Reversed net

The reversed net is the net obtained by reversing the direction of each arc in the original Petri net. Since the assembly and the disassembly operations are assumed to be invertible each other, the reversed net of a disassembly Petri net is the assembly Petri net for the same product.[73,74] By reversing a net, deadlocks and traps exchange roles,[80] which is not a problem in this case, since P-semiflows are both deadlocks and traps. The resulting Petri net is not a free-choice net anymore, but it is still live and safe and has the same P- and T-semiflows as that of the original net.

(5) Brother net

When an AND/OR graph is mapped into an ordinary Petri net, each reversible AND-arc will be decomposed into two transitions which are in the opposite directions.[69,89] These are called brother transitions, t_i and $\bar{t}_i$. Accordingly, if a

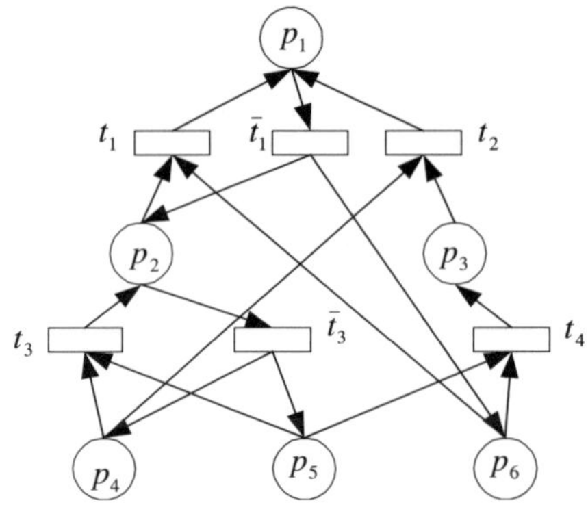

Fig. 17. An example assembly/disassembly brother Petri net.

sequence S is $t_1 \rightarrow t_2 \cdots \rightarrow t_l$, then the brother sequence of S, also written as $\overline{S}$, is $\bar{t}_l \rightarrow \bar{t}_{l-1} \rightarrow \cdots \bar{t}_1$. An example brother Petri net for assembly/disassembly sequence representation is shown in Fig. 17. Assembly sequences are: $t_4 \rightarrow t_2$ ($p_5 \rightarrow p_6 \rightarrow p_4$, $p_6 \rightarrow p_5 \rightarrow p_4$), $t_3 \rightarrow t_1$ ($p_4 \rightarrow p_5 \rightarrow p_6$, $p_5 \rightarrow p_4 \rightarrow p_6$); and the disassembly sequences are: $\bar{t}_1 \rightarrow \bar{t}_3$ ($p_6 \rightarrow p_5 \rightarrow p_4$, $p_6 \rightarrow p_4 \rightarrow p_5$).

4.2.3. *Assembly sequence computing and searching*

To search for feasible assembly/disassembly sequences, two possible requirements may be proposed for the practical implementation situations: one is the problem of searching all possible sequences for the given initial state and the final state, and the other is the problem of searching the optimal sequence under certain evaluation criteria such as assembly/disassembly time, cost, number of steps, or flexibility. The complexity of the first problem in terms of time and space is much greater than that of the second one. The search and control strategies of Petri net can apply the depth-first search method and breadth-first search method. If the optimal planning is expected, the AO* algorithm can be used.[92] The AO* algorithm consists of two major operations, namely a top-down graph expanding procedure and a bottom-up cost revision procedure. Assembly sequences can also be selected by using linear programming techniques.[78,79]

A disassembly Petri net derived from its AND/OR graph expresses all the possible assembly/disassembly sequences for a given product. Thus, sequences can be searched from the Petri net. The planner is not only interested in whether a final state can be reached from the initial state, but he also request the sequence used to reach the final state. The reachability graph (tree) of the net is computed and then a search can be performed for the computation of assembly sequences.[89,91] The number of leaves in the tree is the number of all possible task sequences. The depth of the tree is the length of the sequence. The length of the shortest path from the root to a leaf is the number of operations in the optimal sequence in the sense of number of steps. Either a sequence of transitions, or a sequence of system states can be shown in the reachability tree, that is, the feasible components or operations sequences can be directly found from the reachability tree. Its representation size is much greater because of the many duplicate nodes. However, this procedure has been proven to be very expressive for complex assemblies. It is not economical to find an optimal sequence.

Petri nets intrinsically provide the concept of cyclical behavior and sequences by means of T-semiflows.[80] T-semiflows express the firing count of the firing sequences of a Petri net. This will lead to the same count of the firing sequences of a Petri net resulting in the same marking. Since a disassembly Petri net T-semiflows are boolean vectors, the identification of the actual firing sequence corresponding to a T-semiflow is straightforward. Therefore, the assembly sequences for a disassembly Petri net can be represented by its T-semiflows. The computation of T-semiflows is a standard operation performed by Petri net analysis tools.[81] The computation of a single semiflow is a polynomial operation, although in the general case there might be a generator set (number of semiflows) exponential in the number of nodes (places and transitions). However, there might be an exponential number of them in identifying all assembly sequences. In practice, the computation of T-semiflows is an easy operation (usually several orders of magnitude faster than a state space analysis), since the net incorporates a number of precedence constraints leading to a reduced number of cyclical behaviors. For example, the flashlight in Fig. 16 has ten linear assembly plans, out of the $4! = 24$ available if there were no constraints (permutations in the number of parts). These ten linear plans are expressed by eight T-semiflows as mentioned above.

4.3. *Assembly sequence evaluation and selection*

In spite of the above mentioned constraints, a product may possess quite a large number of feasible assembly or disassembly sequences. In order to select the optimal assembly sequence, it is essential to develop some procedures to reduce the sequence count. The assembly sequence generation algorithm, used for generating the set of all feasible assembly sequences, has focused almost exclusively on the hard constraints imposed by the geometry of the constitute parts and the product itself. In practice, the human planners implicitly consider various criteria or some additional constraints called soft constraints (e.g. cost constraints) which can further limit the available sequence alternatives or enable them to assess the performance of a particular sequence.

4.3.1. *Assemblability analysis and evaluation*

The principles of design for assembly (DFA) involve minimizing the cost of assembly within the constraints imposed by the requirements in order to meet the appearance and functionality of the product being assembled.[1] Analysis of assembly properties of a product is needed during the initial design stage to identify potential assembly problems. An effective and efficient evaluation method plays a crucial role in design by indicating the cause of design weakness through identifying the tolerances, form features, and geometries of assembly parts, rather than simply providing an evaluation score for the assembly parts or assembly operations. Assemblability analysis and evaluation are the keys to assembly design, assembly operational analysis and assembly sequence planning.

The factors that affect the assemblability are classified into two categories: geometry-based parameters and non-geometric parameters. Four characteristic types of the parts and operations involved are of significance: geometry characteristics (related to parts' geometry), physical characteristics, connection characteristics (related to the type of contact between the components), and operation characteristics.[38,57] Since many factors are involved, a multi-order (2-order in this research) model is required to rank them. The first-order factors set is described as: geometric factor, physical factor, connection factor, and operation factor. The second-order factors set can be described as: α symmetry,[1] β symmetry,[1] number of ease of assembly form features, size, weight, fit type, position, orientation, translation, rotation, force/torque, etc. From the model proposed in Ref. 89, assemblability evaluation is based on the additive aggregation of the degree of difficulty of assembly operations. It can be accomplished by evaluating the assemblability of a joint. Two types of joint are considered: the fastener joint (with agent, e.g. screw, pin, bolt and nut) and the operation joint (without agent).

For an operational joint, the secondary part is mated into the primary part, and the assembly difficulty score (AEI) is calculated using the following equation[89]:

$$\mathrm{AEI}(J) = \frac{1}{100}\sum_{i=1}^{n} ds_i(x_i), \tag{1}$$

where $ds_i(x_i)$ is the relative difficulty score of the joint for the ith assembly factor. $\mathrm{AEI}(J)$ is the assembly difficulty score of Joint J, which is regarded as an assemblability evaluation index of Joint J.

For a fastener joint, the primary part and secondary parts are mated together first, and then the agent(s) is used to join the mated parts. Assuming all the assembly characteristics among the mated parts and the agents are equally important, the assembly score for a fastener joint is calculated as follows:[89]

$$\mathrm{AEI}(J) = \frac{1}{100p}\sum_{j=1}^{p}\left(\sum_{i=1}^{n}[ds_i(x_i)]_j\right), \tag{2}$$

where p is the total number of secondary parts and agents involved in the fastener joint. After all the joints are evaluated, the total assembly difficulty scores can be obtained by summing up all the evaluated scores of these joints. As different assembly sequences require different assembly operations, the total assembly difficulty score is therefore different. Figure 18 evaluates a joint operation (handle

	Assembly factor	U_{11}	U_{12}	U_{13}	U_{21}	U_{22}	U_{31}	U_{41}	U_{42}	U_{43}	U_{44}	U_{45}
Mating	Data	360	0	2	5 cm	21b	LN	Clear	Vertical	4 cm	0	0
Operation	Score	4.66	0.59	1.75	0.64	0.40	4.5	0.0	1.67	0.60	0.0	0.0
	Total assembly difficulty score = 14.81, assembly evaluation index = 0.15											

U_{11}: α-symmetry
U_{12}: β-symmetry
U_{13}: Number of ease of assembly form features
U_{21}: Size
U_{22}: Weight
U_{31}: Fit-type
U_{41}: Position
U_{42}: Orientation
U_{43}: Translation
U_{44}: Rotation
U_{45}: Force/torque

Fig. 18. Evaluation data of the design of optic lens.

to receptacle) for the design of flashlight in the sequence: Cap(C) $\rightarrow$ Stick(S) $\rightarrow$ Receptacle(R) $\rightarrow$ Handle(H). The total assembly difficulty score and assemblability evaluation index of this operation are 14.81 and 0.1481 (0.15), respectively.

4.3.2. *Assembly sequence evaluation factors*

For the econo-technical justification, two major category factors: qualitative and quantitative, have been specified and used in Ref. 91. The following one or more important qualitative criteria should be considered for the selection of sequences[52]: frequency of direction changes, stability and security of subassemblies, parallelism among assembly operations, union and modularity of subassemblies, type of assembly cell, and clustering of parts. The detailed description of above criteria can be found in Refs. 3, 6 & 51. The relative importance of the criteria depends on the method of assembly being used. The qualitative criteria pertain to the characteristics or the attributes of the particular assembly states or state transitions (assembly tasks). Particular states of assembly can be either desirable or undesirable from a manufacturing standpoint, and these can be applied in sequence selection.

On the other hand, as mentioned above, although the qualitative criteria are helpful references for the assembly sequence selection, it is inconvenient or difficult to utilize them in practice. The quantitative evaluation factors employ more concrete characteristics about the assembly processes. They are most often associated with the attributes that directly influence the assembly cost. The quantitative characterizations may include the time necessary to accomplish the assembly tasks, the costs of the hardware required, the costs of fixturing or tooling needed to secure unstable states, and so forth. Specifically, the following one or more factors may be potentially considered during evaluation: total assembly time or cost, number of product re-orientations during assembly, number of fixtures, number of operators, number of robot grippers, insertion priority for the specific part, and number of stations.[51,91]

Based on the incorporation and integration with assemblability analysis and evaluation, as well as the predetermined time standard analysis, the following quantitative factors can be reduced: the total assembly time, the cost constraints, the priority index which represents the insertion priority for the specific part, and the number of assembly stations.[91] The reason that these factors are present is that other factors such as the number of product re-orientations during assembly, the number or cost of fixtures and the number or cost of robot grippers are considered directly or indirectly in the assemblability analysis. The cost constraint can be described by predicates such as *time_consuming* and *cost_effective.* The assembly time and cost for each subassembly operation can be estimated and will be assigned to the corresponding transition in the Petri net. Then the final sequence will be the one with the minimum total assembly time or cost. The number of assembly stations can be estimated by the assemblability analysis. In addition it can also be estimated by the user from the number of feasible subassemblies or the number of operations involving one or several base parts in the assembly line. For a part (or a subassembly) priority index, I_p, it is based on the partial precedence constraints as discussed above, as well as the assembly sequence graph. It may be calculated as follows: $I_p = N_n/N_t$, where N_t represents the total number of nodes of the assembly sequence graph, and N_n represents the number of nodes that may be assembled after the given part. The global factor for the whole graph is the average value of all the priority indexes of the part.

4.3.3. *Assembly sequence evaluating and selecting*

Once the feasible assembly sequences have been identified, they can be individually analyzed according to some cost or performance criterion. A simple criterion can be applied to the selection of assembly sequence, such as the shortest time or the lowest cost path through the weighted assembly sequence Petri net graph. The selection of sequences from an assembly Petri net can be done by either deleting the unwanted assembly states and unwanted assembly tasks or by retaining the most desirable assembly states and tasks.[91] From a Petri net viewpoint, each T-semiflow identifies a strongly-connected marked graph component, such that the original disassembly Petri net is obtained by the composition of the marked graphs.

Techniques for marked graph analysis are described in Refs. 80 & 82. If a cost or a time delay coefficient is assigned to each assembly operation, it is easy to compute the minimum cost/delay assembly plan by examining all marked graphs derived from T-semiflows (i.e. sequence by sequence). Delay coefficients divide the marked graph into timed transitions. Note that no assumption is required about the statistical characteristics of the delay parameters. They can be deterministic or stochastic variables. In the latter case the average is considered. For timed marked graphs, each directed circuit C_k yields a minimal-support P-semiflow.

(1) Minimum cycle time

Let τ_i be the delay associated to transition t_i. The minimum cycle time of an marked graph is given by Refs. 76 & 82.

$$\Gamma = \max_k\{(\Sigma_{i\in C_k}\tau_i)/\mathrm{M}_0(C_k)\}, \tag{3}$$

where $\mathrm{M}_0(C_k)$ denotes the number of tokens in C_k in marking M_0. In the marked graphs generated by the disassembly Petri nets, $\mathrm{M}_0(C_k) = 1$, $\forall k$, thus the formula reduces the delay accumulations in each circuit.

Let S_k be the assembly plan associated with the marked graph generated by the T-semiflows TS_k of the flashlight disassembly Petri net. We compute the minimum cycle time for each feasible assembly sequence. The following cost or delay values are obtained by predefined task time analysis and assemblability analysis,[1] as obtained from Ref. 89:

$$\tau_1 = 4,\ \tau_2 = 1,\ \tau_3 = 2,\ \tau_4 = 4,\ \tau_5 = 1,$$
$$\tau_6 = 2,\ \tau_7 = 7,\ \tau_8 = 6,\ \tau_9 = 6,\ \tau_{10} = 2,$$
$$\tau_{11} = 7,\ \tau_{12} = 5,\ \tau_{13} = 7,\ \tau_{14} = 5,\ \tau_{15} = 7.$$

Transition t_{loop} only plays a syntactic role and introduces no delay. The minimum cycle time, each computed for the marked graph associated with one of the T-semiflows, are the following[76]:

$$\Gamma_1 = 13,\ \Gamma_2 = 13,\ \Gamma_3 = 11,\ \Gamma_4 = 13,$$
$$\Gamma_5 = 11,\ \Gamma_6 = 11,\ \Gamma_7 = 13,\ \Gamma_8 = 11.$$

Thus, in this case there are four assembly plans with the same minimal duration. Note that TS_5 and TS_8 define plans whose minimum cycle time is achieved with some concurrency between operations. Alternatively, each of them corresponds to two linear plans with different interleaving, whereas the remaining T-semilflows correspond to a linear plan each. The cycle times above are the same as those computed in Ref. 89 except for the plans S_5 and S_8 which have potential for parallel execution.

(2) Sequence evaluation index

Once the mating operation is evaluated, the entire sequence of operations can be evaluated as well. The evaluation of the entire sequence, requires the comparison and the selection of a preferred one. Therefore the aggregate measure of difficulty for the entire sequence is represented as a fuzzy number between 0 and 1.

Suppose that the following notation is used: S_i = sequence i, $i = 1, \ldots, n$; n_i = number of operations in sequence S_i, S_{ij} = operation j in sequence i, $j = 1, \ldots, n_i$; ds_{ij} = assembly difficulty score that represents the degree of difficulty of operation j in sequence i. For the entire sequence, the assembly difficulty score for the sequence

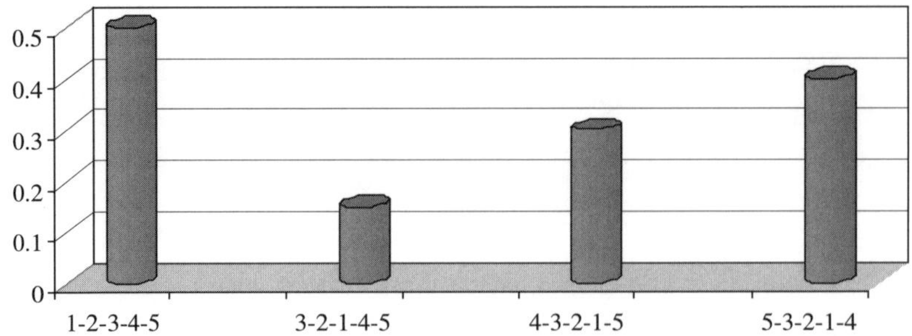

Fig. 19. Assembly difficulty score under different assembly sequences.

i is calculated using the following equation[89]:

$$\mathrm{SEI}(S_i) = \frac{1}{100n_i}\left\{\sum_{i=1}^{n_{i_1}} ds_i(x_i) + \frac{1}{p}\sum_{k=1}^{p}\left(\sum_{j=1}^{n_i - n_{i_1}} [ds_j(x_j)]_k\right)\right\}, \tag{4}$$

where $\mathrm{SEI}(S_i)$ is the sequence evaluation index which is represented by the assembly difficulty score; $n_i = n_{i_1} + n_{i_2}$, n_{i_1} is the number of operational joints in sequence i, and n_{i_2} is the number of fastener joints in sequence i; $ds_i(x_i)$ is the relative difficulty score of the joint for the ith assembly factor; p is the total number of secondary parts and agents involved in the fastener joint. Based on Eq. (4), the preferred sequence is the one with the lowest SEI.

To be in line with the fuzzy evaluation of assemblability, the fuzzy evaluation of assembly difficulty score is carried out for assembly sequences. For the optic lens assembly (Fig. 25), the effects of four different assembly sequences ($O_1 \to O_2 \to O_3 \to O_4 \to O_5$, $O_3 \to O_2 \to O_1 \to O_4 \to O_5$, $O_4 \to O_3 \to O_2 \to O_1 \to O_5$, $O_5 \to O_3 \to O_2 \to O_1 \to O_4$) on assembly difficulty score are shown in Fig. 19. Among these four assembly sequences, the optimal one with the minimum assembly difficulty score is $O_3 \to O_2 \to O_1 \to O_4 \to O_5$.

4.4. *Assembly sequence simulation and animation*

The initial marking M_0 of the Petri net adds tokens to places in the initial states of an assembly to trigger the simulation. The goal of the assembly (sub) Petri net graph and the reachability tree simulation is to visualize and animate the assembly sequence and its executive processes. With this simulation functionality, the user can observe the execution of assembly processes. The following two search and control strategies for Petri net simulation and animation are hybridly used: the concurrent and asynchronous event dispatching method and the continuous transition scanning method.[92] A subassembly and an assembly operation are regarded as events, and so is a part arriving at an assembly station or a buffer station waiting for assembly. All these events can fire the Petri net asynchronously, but the whole system can be carried out concurrently. On the other hand, after an event fires an assembly system

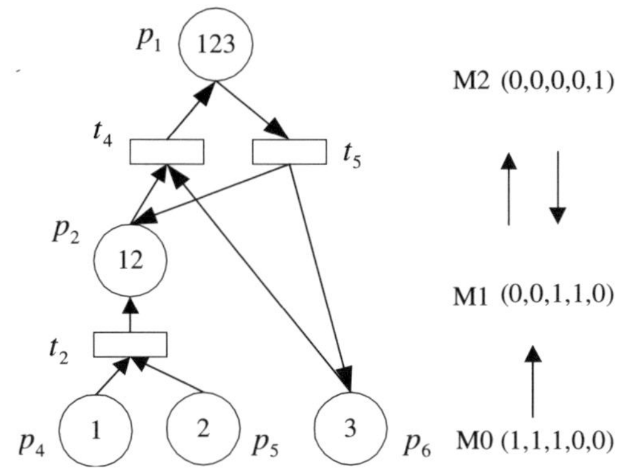

Fig. 20. Sub graph and reachability tree.

to carry out a function, the Petri net continuously scans the places and transitions, and checks whether the firing conditions of a transition are satisfied. If the firing conditions of a transition are satisfied, it will be fired, and tokens of the system will be changed. The system will run continuously until there is no more transitions to be fired.

The number and distribution of tokens in places denote the dynamic system states in the executive processes. As the data structure of Petri net is an incident chain-list, the execution of the Petri net is implemented by treating the data in its corresponding nodes. When a transition is fired, the corresponding data block is marked. After the execution is finished, the incident chain-list of the sub Petri net is obtained by searching the data nodes of the Petri net from top to bottom, and then the reachability tree can be calculated and visualized. In this simulation, the shortest sequence is defined as the sequence of the minimum fuzzy difficulty score (or minimum number of steps or the minimum total assembly time). Figure 20 shows an example of a sub Petri net and a reachability tree. For most applications, more than one feasible sequence may be generated and an evaluation and selection strategy is used to select the best one. Normally, the searching sequences from higher level nets may be useful to verify the correctness of decomposition and the final net is searched to generate a final task sequence.

4.5. *Algorithm of automatic assembly sequence planning*

Generally, the automatic generation process of sequence will experience the following steps:

Step 1: Program starts; Input the primitive n_part_list , $[p_1, p_2, \ldots, p_n]$;
Step 2: If n_part_list is null, then there is no solution. Go to Step (10);
Step 3: Create or load assembly model (solid model of a product, features, P/T net and assembly incident matrix, some constraints); Assembly constraints and precedence relation (directed P/T net) is created;

Step 4: Create level i ($i = 0, 1, \ldots, n-1$) subassemblies and construct a corresponding list;

Step 5: Select a subassembly (one at a time) from the level i subassembly list to test whether the assembly constraints satisfy by geometric reasoning and calculation; Obtain all level i feasible subassemblies and then construct a list;

Step 6: Decompose each of the level i feasible subassemblies into assembly operations, and the remove all unfeasible subassembly operations; Construct a list of all feasible subassemblies;

Step 7: Generate the assembly operation sequences and obtain part chain-list of assembly sequences; Draw the assembly sequence Petri net graph;

Step 8: Evaluate and select the assembly sequences by the way of econo-technical justification with qualitative and quantitative criteria (e.g. assembly time and cost, assembly operation difficulty score); Iterative refinement of Petri net graph;

Step 9: Petri net simulation and animation; 3D assembly animation;

Step 10: End.

5. Integrated Knowledge-based Assembly Planning System

As discussed above, all assembly or disassembly sequences of a product can be generated in an efficient way. However, this method needs much more reasoning experience or intervention and it is so inconvenient that the user has to verify as many feasible sequences as $n!$ (maximum) for an n-part product. It is too complex and time-consuming to cope with manually, especially for products with many more parts. An knowledge-based assembly planning system is required to generate, select and evaluate the assembly or disassembly sequences automatically. Together with the developed assembly modeling and representation, the knowledge-based system can be designed to identify these precedence constraints based on geometric information and then deduce the geometric relationship of the assembled parts.

This section demonstrates the effective use of an integrated knowledge-based approach in solving the problem of formulating assembly plans for general mechanical assemblies. The implementation of the integrated knowledge-based assembly planning system (IKAPS) will be presented below.

5.1. *System overview and architecture*

The developed assembly planning system was implemented using C/C++ and the CLIPS (C Language Integrated Production System) expert system development shell. All recommendations about product design for assembly, and assembly sequences including optimal assembly sequence can be generated and visualized by automatic reasoning, searching, computation and simulation from a comprehensive assembly knowledge base when the user answers the questions presented by the system interactively. It exhibits superiority over existing systems in the following aspects:

(1) It integrates the assembly design, the assembly evaluation and analysis, and the assembly planning process simultaneously;
(2) It permits a multitude of various component assembly trajectories;
(3) It incorporates a robust part collision detection algorithm;
(4) It is able to generate and optimize the assembly strategies (plan and operations); and
(5) It visualizes design and assembly process in a quick and intuitive manner.

The optic lens assembly (Fig. 25) will be used to illustrates the effectiveness of the developed system. Based on the assembly application framework and the algorithms of design and planning as discussed above, IKAPS is implemented in an integrated knowledge-based environment. Figure 21 is a block diagram illustrating the integrated intelligent CAAP system concept. As illustrated in the figure, the IKAPS environment consists essentially of six modules, which integrates the assembly sequence generation with CAD, geometric modeling and reasoning, assemblability analysis and evaluation, and predetermined time standard analysis. A graphical interface with multiple independent software modules provide feedback to the user while performing a design-and-planning related task. The integration approach is based on the data and the knowledge-driven, modularized structures as in Ref. 91:

(1) Module 1 (AMD): assembly modeling and design;
(2) Module 2 (CE-ES): concurrent engineering (CE) expert system (ES);
(3) Module 3 (DFA): design for assembly (DFA);
(4) Module 4 (AP-PN): assembly planning (AP) and Petri net (PN) modeling;
(5) Module 5 (ATAE): assembly task analysis and evaluation;
(6) Module 6 (ASA): assembly simulation and animation.

Figure 21 illustrates the structure and the information flow of the knowledge-based assembly system from a top-level, modular perspective. The system is supplied with the assembled product in a boundary representation (B-rep) format. It typically include an efficient user interface to facilitate the task of representing and manipulating assemblies and their components. The AMD module is used for assembly modeling and design by incorporating the "place-transition" structure modeling, feature-based and geometric solid modeling techniques. The assembly editor (submodule of design module) can also accept imported CAD files of individual components from DXF-based modeling systems and organizes them into an assembly representation. Using feature recognition techniques, the assembly editor can differentiate the joints from the parts and the assembly features on the individual parts. The DFA(AE) module is for the assemblability evaluation and is designed for assembly. The output of this module which is the assemblability evaluation index,

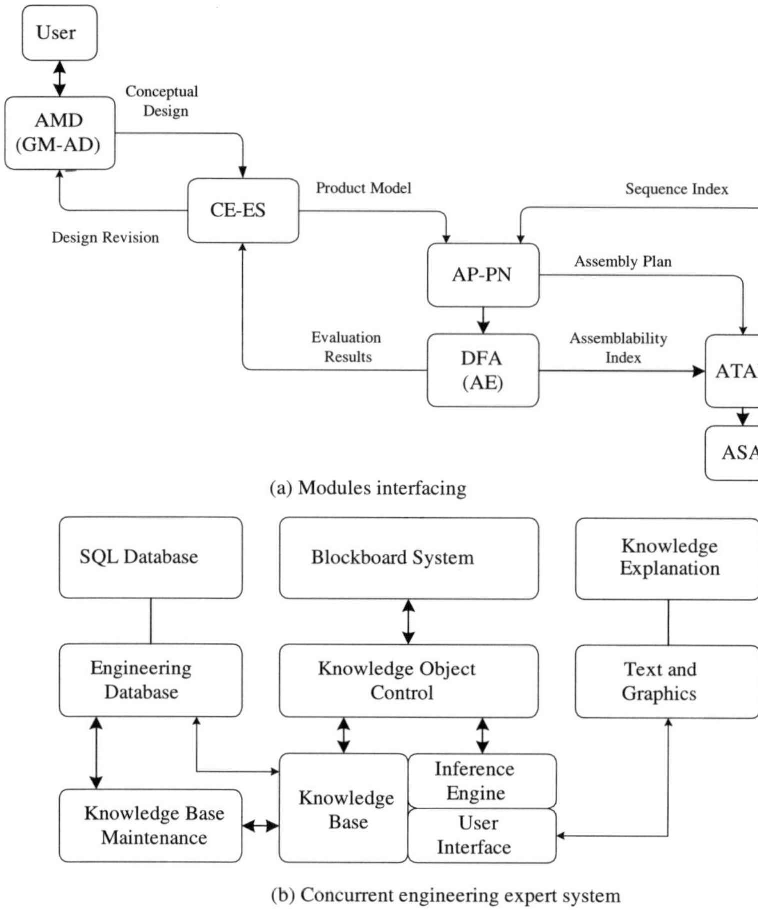

Fig. 21. IKAPS assembly system architecture.

is used to assemble task analysis and evaluation, and to design revision suggestions to the expert system (CE-ES) and the assembly design module for change. The AP-PN module allows users to generate and synthesize assembly plans interactively. The assembly sequences (i.e. macro plans) are generated, evaluated and optimized automatically, and can be automatically converted into low level operation sequences and part motions (i.e. micro plans). The ATAE module allows for the assembly task analysis and evaluation, the assembly operation predefined time analysis, the assembly sequence analysis and the evaluation to be carried out. Using the assembly simulator and animator in the ASA module, the users select and control various simulations (such as interference and tool accessibility). The animation viewer allows the assembly operators to view the modeled assembly process interactively. The users can also randomly access any particular operation in the assembly sequence and then interactively change their 3D viewpoints. The CE-ES module is built upon the CLIPS expert system shell, which has six major components (submodules), namely knowledge-base, knowledge maintenance, inference engine, blockboard, control, and user interface. The knowledge-base consist of database, static knowledge base and dynamic knowledge base. The inference engine uses the knowledge to arrive at conclusions and resolve the rule conflicts. The knowledge maintenance module is used either to check the consistency of knowledge, or to modify and append the knowledge. When new knowledge is added, replaced, or deleted, it can be used to check the consistency between the new knowledge base and the old knowledge base. The control module is used to control and start the system. In addition, it is used for changing the running environment, as well as making decisions. The blackboard is a dynamic storage region used for storing common information and intermediate results and also, exchanging information among modules. It is a shared and structured database that allows the modules or subsystems to access the necessary information and interaction autonomously. Each knowledge source has continual access to the state of problems on the blockboard. With such an integrative framework, the appropriate information or data can be applied smoothly for rapid concurrent assembly design and planning.

Typically, these system components can be used in the following scenario. A designer creates an assembly design using the assembly design and modeling functions in IKAPS. The designer then uses the planning module to generate and select the assembly sequence. Thereafter, the designer can select the simulation module to compose a customized simulation. Based on the simulation feedback the designer may need to refine the assembly design. After several design iterations, the designer is satisfied with the design and then hands it over to the process engineer. In parallel, using the construction module of the workplace and assembly system, the process engineer has created a model of the workcell in which this assembly will be performed. After incorporating the assembly in the workplace, the process engineer performs a detailed simulation to check for potential problems in the final assembly plan. The designer then generates an animation of the assembly process which is downloaded to the operator's or robot's desktop computer where it can be viewed by the operator using the animation viewer. The operator or robot can then start assembling the parts immediately, without the need for extensive training or tedious documentation.

5.2. *System implementation*

The desired system functions have partially been implemented through some well developed sub-systems: design, planning, evaluation, simulation and animation. In what follows, the attention is paid to some specific techniques for the system implementation in an integrated knowledge intensive environment.

5.2.1. *Programming language*

The software development environment includes programming language, development toolkit, and its operating system. Due to the complexity of the knowledge intensive system, the selection of the programming language must take into account factors such as availability, speed, efficiency, ease of use, ease of knowledge representation, compatibility and importability. Information processing in machine and process design is inherently model-based, because the design object is structured in type. Therefore, object-oriented (O-O) programming language would be desirable for knowledge representation. The class and instances mechanism in O-O technology can handle with ease the relationship between objects such as prototype physical structures and embodiment physical structures. The inheritance, encapsulation and polymorphism features of the O-O technology will offer a great flexibility in organizing the hierarchical networked design and planning information structure. However, for the initial implementation of IKAPS, multiple languages (C, CLIPS, COOL) are used to code the design and planning procedures in a hybrid environment. The operating system environment could be any graphical windows-based environment on the PC platform.

5.2.2. *Implementational views and control*

To develop IKAPS, four main tools are used: Windows, which controls the pieces of the user-interface; C, in which all the system control functions, the design and planning output, and the mathematical functions are written; CLIPS,[70] in which the expert systems are written; and COOL, the object system part of CLIPS, in which the design and planning database is created. The relationship of these four tools to the user is shown in Fig. 22. The user interacts through Windows widgets to model the assembly and to produce design changes which are handled by the C code. These changes are then asserted as facts to CLIPS, which runs the rules and outputs decisions by working back up the control chain to the interface. The C code can act directly on the objects in the design and planning stages by communicating with COOL. COOL can be controlled from within CLIPS. The greatest responsibility of the C code is to provide a link between Windows and CLIPS. It converts the user's action into facts that can be asserted in CLIPS, or into changes made to object instances in COOL. CLIPS developed modules or submodules which are able to output advices to the user via Windows using the C functions. The C code and the CLIPS code are described, both asserting facts in the form:

```
(designfact (time ?value) (level ?value) (aspect ?value)
            (module ?value) (information ?value))
```

Figure 22 shows the logical version of the control and data flow between modules in the IKAPS. In the implementation, the modules or submodules entitled *geometric modeling, P/T modeling, graphics, assembly simulation and animation, etc.* are all implemented in C. Windows shares the blocks entitled *user* with the actual user of the system. Concurrent engineering expert systems including *blackboard system, knowledge object control, user interface, knowledge explanation,*

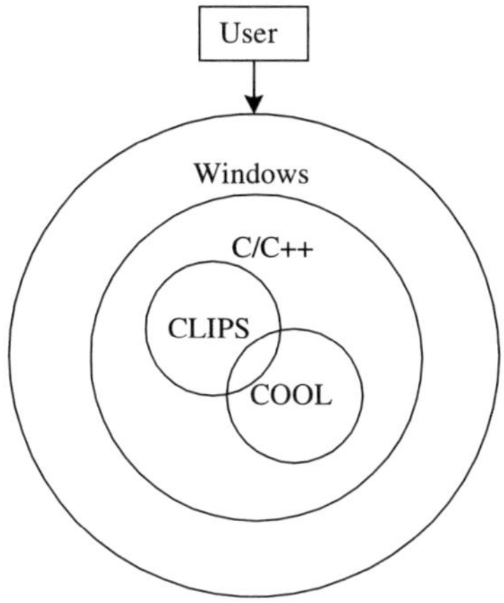

Fig. 22. Implementational view of control and data flow.

assemblability evaluation, *assembly task analysis and evaluation*, and *assembly planning* are implemented by CLIPS entirely. The assembly design and planning database is implemented in COOL. For example, the screw class and snap class can be illustrated as follows:

```
(defclass SCREW (is-a CONNECTOR)
        (slot type (composite)(default SCREW))
        (slot cost (composite)(default 1))
        (slot assembly-time (composite)(default 3))
        (slot disposal-cost (composite)(default 1))
        (slot illegal-materials (composite))
        (slot angles-allowed (composite)(default 0 45 90)))

(defclass SNAP (is-a CONNECTOR)
        (slot type (shared)(read-only)(default SNAP))
        (slot cost (composite)(default 2))
        (slot assembly-time (composite)(default 1))
        (slot disposal-cost (composite)(default 1))
        (slot illegal-materials (composite)(default STEEL))
        (slot angles-allowed (composite)(default 0 90)))
```

There is a common fact list that is controlled by CLIPS, which allows all knowledge to be seen by the separate modules. The fact list also allows the modules to communicate with each other by posting to the fact-list. The "critic" posts criticisms to the fact-list, while the "suggestor" reads these and adds suggestions to the list. Other modules can use the fact-list as a communication center as well. The knowledge-based system incorporates both the declarative and procedural knowledge. The CLIPS inference engine includes truth maintenance, dynamic rule addition, and customizable conflict resolution strategies.

The blackboard system allows the separate knowledge systems to obtain information provided and then either ask for more information from others or supply information to others.[97] It consists of a controller and several knowledge sources. The controller uses its rules to determine which knowledge source will run and when that knowledge source will run. Each knowledge source is also totally independent of the others. They can use different approaches to solve the same problem. Both the logical view and implementation view of the blackboard system are shown in Fig. 23(a) and (b). Each expert system posts information to the blackboard, and then reads the relevant information from it. Each expert system will react if possible. The communication between the user and the blackboard is a logic one only. As shown in Fig. 23, the information to and from the user passes through several other subsystems to achieve the required communication. However, this does not change the view of the system that the user gets when interacting with a blackboard. The user is still able to obtain information from other modules and the post information back to them, just as when they are using a blackboard. The blackboard is used to keep the modules separated at the logical level, so that they can represent different "people". This maintains the view of integration of many different designs and planning modules. Information is placed where everyone can see it and each module is allowed to add any relevant information to it. For the implementation of the blackboard system, it is a combination of the CLIPS fact list and the COOL object list. These lists are available to the user through the C program and Windows, and they are also available to each of modular rules. As such, each of the modules has access to whatever information it may find relevant. The module rules are separated into different rules and are read into the CLIPS shell on program execution. So there are seemingly separate modules, reacting to the same information and providing each other with information from across the different aspects.

5.2.3. *Knowledge acquisition and data integration*

The knowledge of how to design and plan assembly in the IKAPS, and when to invoke other modules, is gained by carefully stepping through the process of designing and planning a specific product. The approach taken in building most existing expert systems requires an expert to dictate rules directly to a knowledge engineer for symbol encoding. Experts usually explain complex concepts by way of examples rather than by stating principles. The knowledge engineer must then form general rules from the way of expert solving problems. It is this library of knowledge and experience that the inference engine draws on to solve problems. Handbooks and books on design for assembly are two of the main sources of knowledge used in building the knowledge base for assemblability analysis and evaluation. Other sources include consultations with production experts in a company or factory. The domain knowledge in IKAPS can be acquired in two ways: one is to call a whole screen editor, and the other is to call an interactive knowledge-based editing tool. The former method is suitable for the input of a large amount of knowledge at

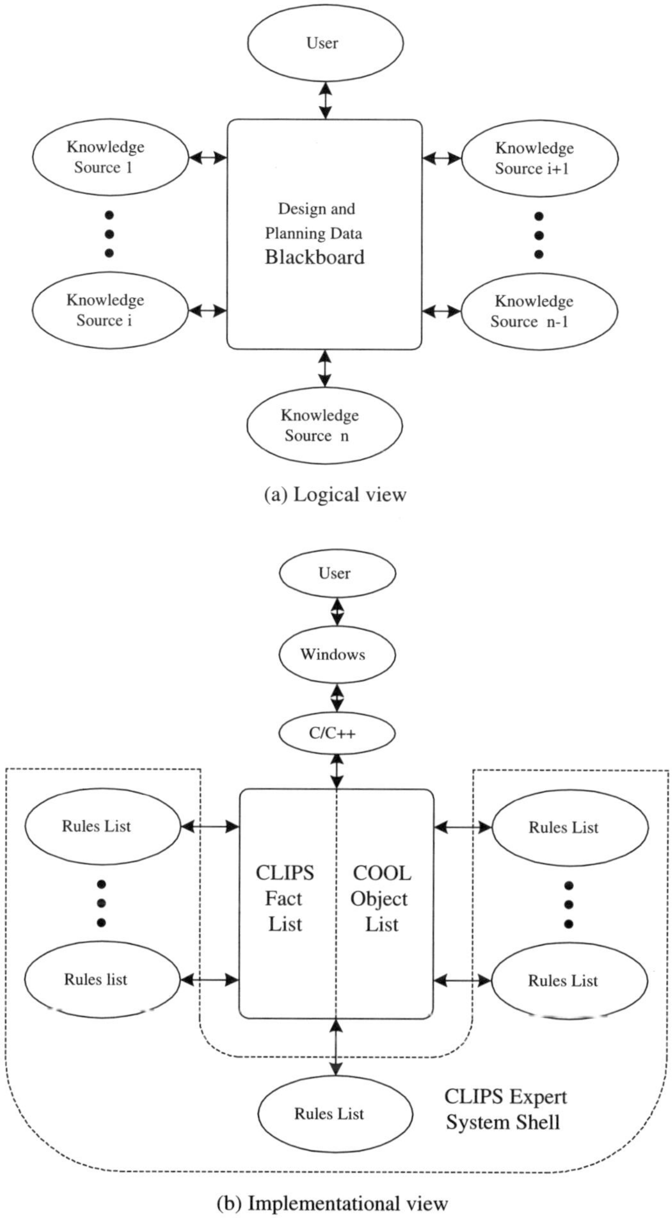

Fig. 23. The blackboard system.

one time, especially in the initial stage of the setup of the expert system. The latter method is suitable for the maintenance of the knowledge base, which provides means for the dynamic modifying (appending, deleting) and testifying of the knowledge. In this way, IKAPS can obtain the new knowledge continuously in the course of application.

Integrating IKAPS with other existing information systems is critical to the success of an assembly automation application. These information systems can include CAD systems, corporate databases, and existing engineering analysis routines. IKAPS includes modules for sharing geometric data with popular CAD packages such as AutoCAD. It can also generate STEP files. AutoCAD can generate DXF files, which is the published format for AutoCAD, from their generative model. The STEP output tools provide the ability to write a STEP file for data exchange. The implementation of STEP supports configuration management information, product structure, wireframe, surface, and advanced B-rep solid geometry. With this interface, the production user can proceed directly from a generative model to the prototype parts. The IKAPS database interface provides access for generative models to SQL database such as Oracle, dBase, and MS Access database.

5.2.4. *Assembly modeling and design module*

The intelligent design module is a knowledge-based modeling system obtained from geometric, feature-based, functional and technological specifications. It incorporates a geometric solid modeler, and a CLIPS expert system development shell. Therefore, this module system is a feature-based, constraint-based, knowledge-based and an object-oriented design environment. Its output is the assembly model. Figure 24 depicts the architecture of the intelligent design module. The geometric modeler (*GeoObj*)[89] is both a solid and region prototype modeling system. External applications designed to interact with *GeoObj* can be linked through an interface. The feature-library in an application programming interface (API) provides functions for creating and manipulating solid components as well as for interrogating and evaluating the geometric and topological properties of solids. CLIPS shell is integrated with the *GeoObj* by rewriting and compiling standard CLIPS input/output routines with *GeoObj* and API functions. The run time of CLIPS can be loaded and executed within *GeoObj*. Therefore, solid objects can be created and analysed by CLIPS applications. To implement the integration of design and assembly, features should contain meaningful information for different application domains. From the assemblability evaluation, assembly reasoning, and assembly process planning viewpoints, each feature is an individual geometric or knowledge entity (e.g. geometric shape). It represents a set of assembly processes and the available assembly tools. The type of feature and the feasibility of the feature instances can be evaluated with the CSG-based solid modeler using Boolean operations and query functions. The traditional exchange of static information by file transfer does not fulfil the requirement in terms of data complexity and speed. Therefore, a dynamic interface is needed for this application. The solid modeler must provide certain facilities to allow other systems to access the geometric core directly.

The feature-based system is tightly coupled with the geometric modeling system (*GeoObj*) in the design module. All the geometric calculations and manipulations

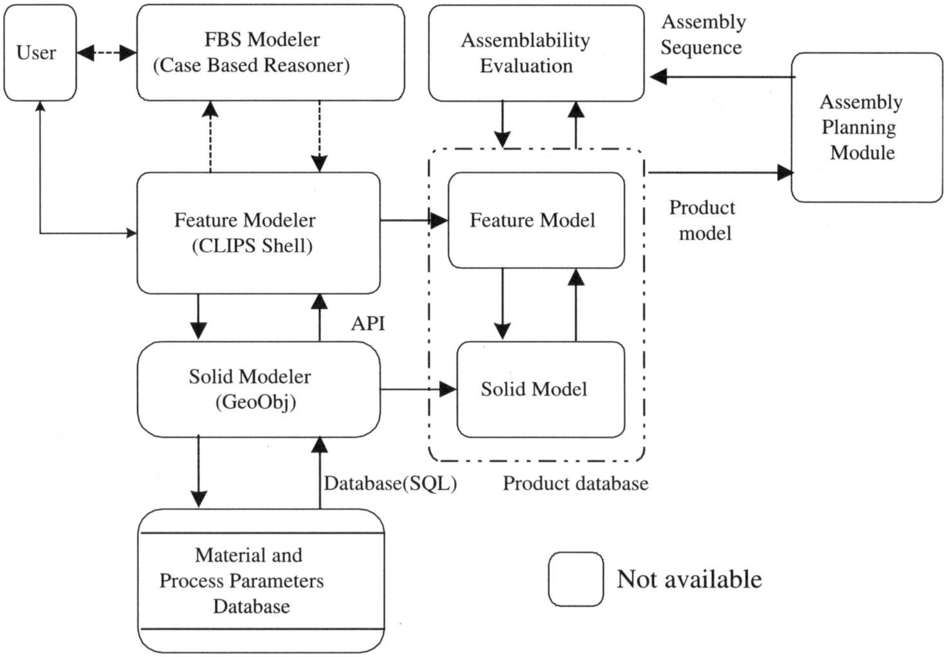

Fig. 24. Intelligent modeling and design module.

are performed through the external procedural calls to the solid modeler within the feature-based system. The subsequent updating of the geometric database gives an immediate feedback to the feature-based system. If the design violates any of the feature constraints the subsequent operation will be rejected and the original model is maintained. These operations are governed by a knowledge-based expert system. The feature library manager provides interactive facilities for the specification of feature classes and for their organization in application-specific feature libraries. A feature library stores its feature class specifications in COOL. These specifications can be loaded into the feature modeler at run-time.

The information from the feature models of all views is represented in a central constraint P/T graph. In addition to the constraint graph, a cellular model is maintained, storing the combined geometry of the feature shapes of all views. The cellular model is, among other things, used for view conversion, feature interaction management, and feature visualization. Information from the 3D design of a product from a CAD package will be input into the system for the assembly sequence planning. Figure 25 shows five parts to be assembled into an optic lens, which are labeled as O_1–doublet 1, O_2–spacer, O_3–doublet 2, O_4–lockring, and O_5–subassembly 1. The subassembly is composed of sp_1, sp_2, sp_3 and sp_4.

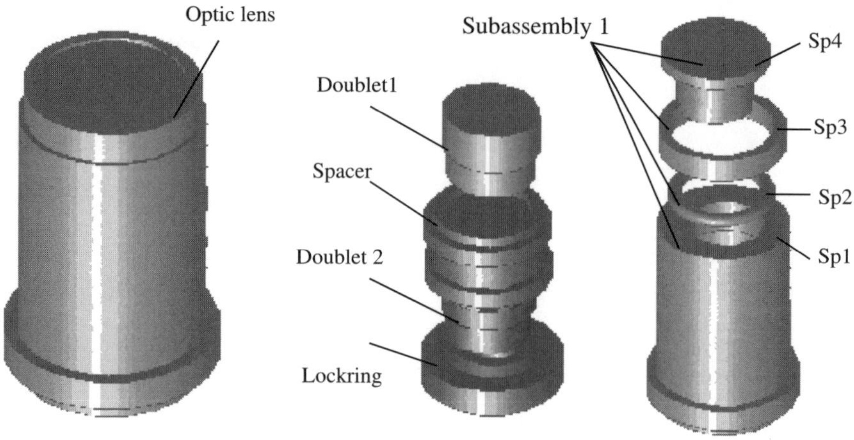

Fig. 25. CAD model of optic lens assembly.

5.2.5. *Assembly plan generation module*

Figure 26 shows a block diagram of an intelligent planning module used for assembly sequence planning. The diagram includes the generation and representation of assembly parts and operation sequences. It incorporates an interference and collision detection module, a sequence generator, and a Petri net tool.

(1) Geometry checking, interference detection and tool accessibility

As described in Sec. 3, interference detection for disassembly operation is required for the assembly sequence generation. It is important to avoid part-part and part-tool interference during assembly sequence planning and assembly operations. The geometry checking employs a collision detection algorithm for every component in every candidate assembly direction. The objective is to determine the set of components which would obstruct the assembly operation if they were already in their final position, or similarly, consider the disassembling of the final product. Under the assumption that all components are rigid and there are no internal forces in the final assembly, the reverse of the disassembly sequence is a valid assembly sequence.

In real-world problems, assemblies tend to contain a large number of parts, each with complicated geometries. For most interference checking algorithms, performance deteriorates rapidly with the increase in the number of parts in the assembly and the number of faces per part. Most assembly design and planning problems exhibit several levels of nested combinatorial explosion. Computation time for detecting interference depends on the size of geometric models. Models can often be simplified before interference checks. In this research, a hierarchical approximation strategy is used for the simplification of the geometric models of individual parts. For instance, the checking for interference between the bounding enclosures of parts can be a quick first test for the presence of part-part interference. Only when the bounding boxes intersect, then it is necessary to proceed with more complex tests.

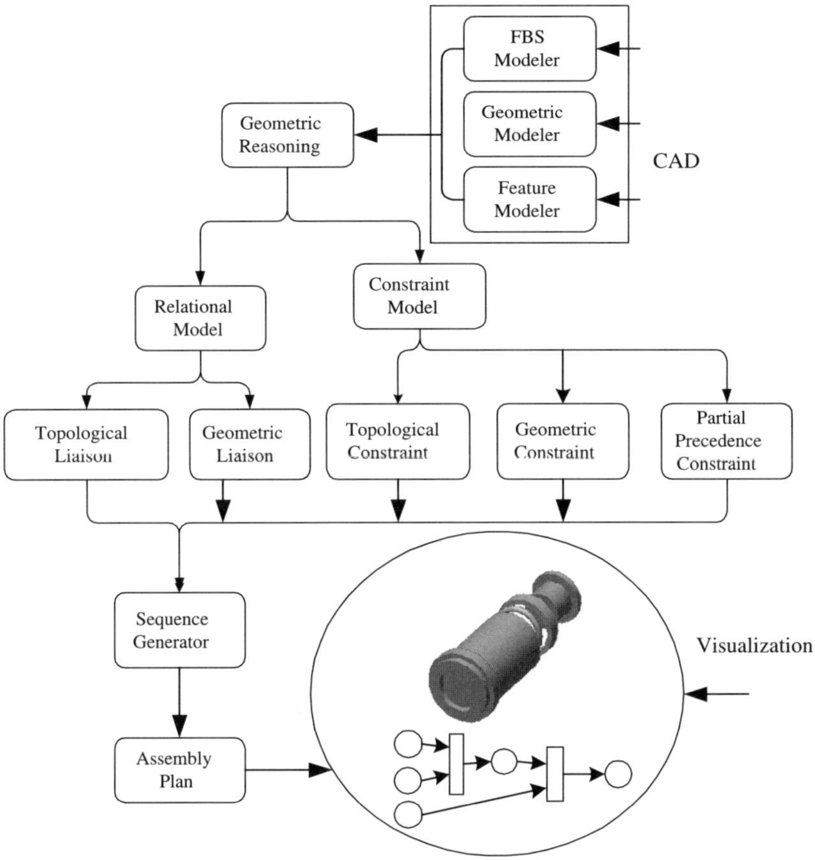

Fig. 26. Intelligent assembly planning module.

This module will determine which components obstruct the removal of a particular component by "graphically" projecting the component in question onto a given plane. The projections of all other components on the same plane which do not have null intersections ascertain the component's obstructions. The output of this module is a list of obstructions which will dictate the precedence relationships of the assembly plan. A sub module has been developed for detecting part-part interference during assembly operations and allowing designers to either modify the design or change the assembly plan to eliminate such problems. Figure 27 shows an example of detecting part-part interference in an optic lens assembly. Designers currently rely on geometric model (virtual prototype) to investigate tool accessibility issues by geometric manipulation.

(2) Precedence generation

The precedence generation module determines the precedence relationships among parts. The output of this module along with a liaison graph defines the product

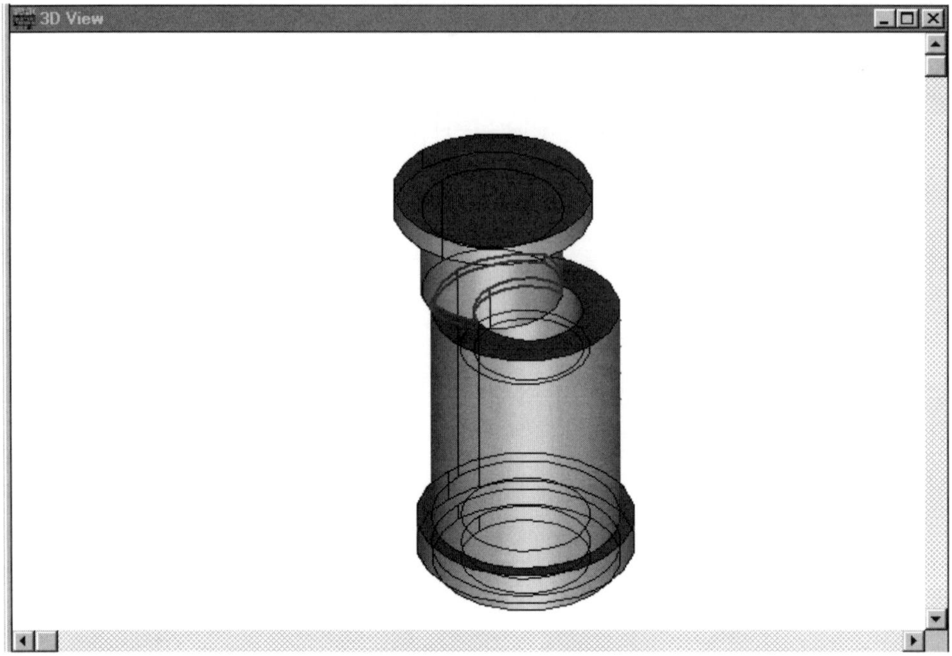

Fig. 27. Part-part interference.

graph. The product graph represents the final assembly where nodes are the components and links correspond to the physical liaisons among the components.

(3) User input constraints

As discussed in Sec. 3, there are many constraints needed to be considered when planning an assembly sequence. Allowing the user to input constraints or criteria on which assembly sequences are chosen helps to prune the number of feasible sequences. The engineer has knowledge about the plant's technology (e.g. the use of special tools) that must be considered in the assembly or other soft constraints such as part stability, personnel safety, etc. Once the soft constraints are identified a more difficult problem would be to quantitatively define them so that they can be incorporated into a computer system. The criteria must be well defined, yet, as the products change, the system must be flexible enough to allow criteria changes without requiring substantial computer input. Once the constraints are quantified, a weighing scheme must be developed so that the most important criteria are given the highest priority during sequence generation and evaluation. The weighing may change as products and plants change. Therefore, it is still necessary to allow engineers to change the importance of the criteria as well as to select the criteria important for that product. The manufacturing engineer's ability to input constraints to the system and understand system output is critical.

(4) Sequence plan generator

Construction of the assembly plan is performed by the sequence generation module via a disassembly approach. At any point in the disassembly sequence, a number of components may be removed from the subassembly. The assembly plan follows directly from the derived disassembly sequence. The submodule is a generic tool that can be embedded in a knowledge-based system like an engine for local constraint satisfaction problems solving. It is composed of three modules:[84] main module, search module, and propagation module. The main module contains the top level loop including the I/O rules. The search module implements a classical chronological backtracking algorithm, while the propagation module contains the forward checking algorithm. To solve any particular problem, the users only need to declare each variable with its domain, including the specific domain description and rules of propagation. All the specific knowledge are encapsulated in the propagation module. For assembly sequence generation, the proposed Petri net representations of assembly planning as discussed in Sec. 4 are incorporated into the main, search, and propagation modules.

(5) Trajectory planning and collision detection

During assembly process, moving a tool from its initial position in the workspace environment to the application position (goal position) requires a sequence of translations/rotations. However, there is currently no convenient mechanism available for entering a 6-DOF path into the computer. It is a goal to relieve the user from the path planning task all together, and the generate the tool and part paths automatically. The 6-DOF path planning is a very challenging problem. Firstly, the tool tends to be in contact with the part to which it is applied in the application position. That means that in the configuration space the goal configuration may be almost completely surrounded by obstacles, which is difficult to handle for path planning algorithms. Secondly, one only computes a path once for each part in the assembly. This means that one cannot amortize any of the pre-computations required by some of the path planning algorithms.

Several path planning algorithm have been developed specifically for assembly planning.[85–87] The implementation in this chapter is based on a group of randomized path planners.[71] These algorithms do not require the computation of the C-space, and they consist of the following components: a simple potential filed method, generation of random via-points, creation of a roadmap graph, and search of the roadmap. A simple potential field method described in Ref. 72 is implemented. The method allows the tool to make minor moves to 36 neighboring positions (6 pure translations, 6 pure rotations, and 24 combined rotations/translations), and it ranks these positions according to their distances from the goal position. The tool is moved to the immediate neighboring position that is closest to the goal without colliding with any obstacles. The algorithm terminates when the goal is reached or when no neighboring positions closer to the goal are collision free.

To check for collisions in the IKAPS system, a fast collision detection algorithm is used based on the oriented bounding-box trees.[71] It uses a faceted representation of the tools and objects, and collisions are detected between objects that are in contact but not intersected. For instance, when rotating a cylindrical object inside a cylindrical hole of the same diameter, this algorithm will always detect a collision except when the facets of the hole happen to line up perfectly with the facets of the cylinder. Since contact situations are common in the assembly, the geometric kernel is used to check for intersections between the exact solid models whenever the algorithm detects a collision between the faceted models. In this way, collisions can be checked rapidly and accurately.

5.2.6. *Assembly task analysis and evaluation module*

The intelligent evaluation module is used for the analysis and evaluation of the design and planning results, the determination of redesign and replan, and the optimization from assemblability and economics. It includes the assembly design evaluation and the planning evaluation sub-modules with both assemblability and econo-technical considerations.

The assembly design evaluation sub-module deals with design evaluations, redesign evaluations, and assemblability evaluations. As discussed in Ref. 89, the assemblability is described by assembly operation difficulty, which can be represented by a fuzzy number between 0 and 1. Due to the complexity and uncertainty of the assembly problem, the assemblability is analyzed and evaluated using fuzzy sets. In Sec. 4.3.1, the assemblability evaluation is outlined for assembly sequence evaluations.

Integration between this sub-module and geometric model (*GeoObj*) is carried out with two geometric feature calculation algorithms. This is shown in Fig. 28, which calculates the overall dimensions and rotational symmetries of the part data. Information from the intelligent design module is first translated into a solid CSG representation scheme before they are processed by the algorithms. The system implementation consists of two main modules: the pre-processor and the assembly analysis module. The pre-processor reads the part model data and the assembly description from the user. The part data is translated into a suitable format for the assembly analysis module. The analysis module contains the two algorithms and analyses all the assembly components as translated by the pre-processor.

The assembly planning evaluation sub-module is used to analyze, justify, and optimize the assembly processes and operations which include assembly sequence, task sequence and task sequence scheduling using fuzzy sets and Petri nets based algorithms.

5.2.7. *Interactive visualization module*

The primary goal of interactive visualization module is to provide or present sufficient information to the user in a clear and organized manner. Previous assembly

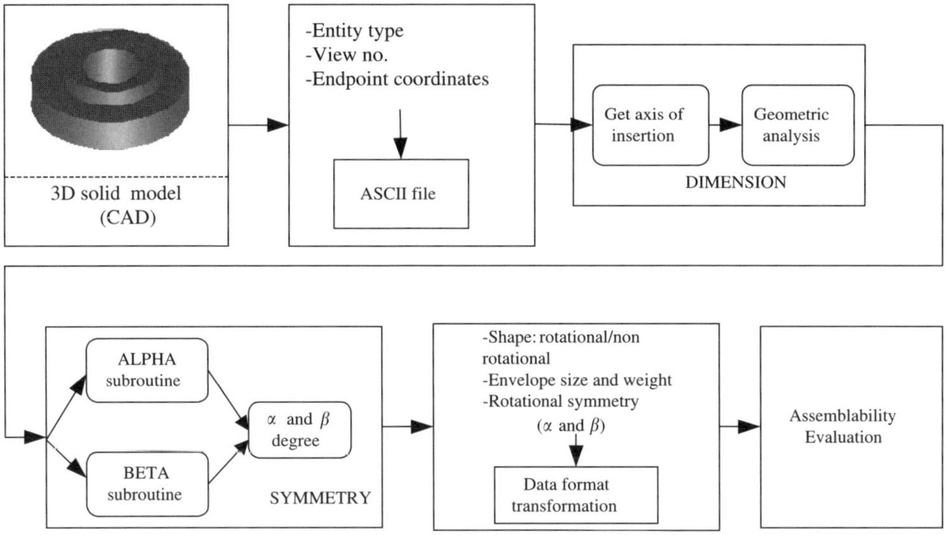

Fig. 28. Integration of DFA and CAD.

planning systems have relied on a single representation and visualization method to facilitate the comparison, evaluation, and selection of desirable sequences. Several factors are usually considered when evaluating a sequence, including the number of reorientations, the number of fixtures needed, the stability of the assembly at various stages, the relationships between various sequences, the insertion priority for specific parts, and the number of assembly stations. Based on these criteria, the visualization method must be capable of presenting information ranging from global details (e.g. the relationships between sequences) to local details (e.g. fixturing and part stability). It is unlikely that a single representation of sequences can adequately provide all of this information to the user.

The developed IKAPS takes a Petri net alternative approach to represent and visualize the assembly sequences. Based on Petri net operation, decomposition, and refinement,[55,89] this approach allows the user to interact with the system to create visualization that are optimum for his/her own viewing, and it provides a technique that can chunk information. There are three general types of visualizations needed during assembly planning[96]: global, comparative, and detailed visualizations, in which the global visualizations provide the user with information about multiple feasible sequences; the comparative visualizations allow the user to quickly compare two sequences; and the detailed visualizations examine the sequences in detail.

(1) Global sequence visualization

To narrow down the number of assembly sequences, abstract visualizations that provide information about the sequences simultaneously are needed. These visualizations should include the ability to group assembly sequences based on the criteria defined by the user. The Petri net techniques allow the user to interact with the system by adding constraints, creating the Petri net graphs representing the feasible sequences, and then creating a node or sequence diagram. The nodes can be grouped based on the criteria that are met and the user can specify what criteria are reviewed. Thereafter he can assign visual attributes to the criteria. For example, the user can specify that the nodes are grouped based on the number of reorientations, and he can also assign the color or size of nodes as attributes. These graphs contain no detailed part information and are used primarily for illustrating the relationships among various assemblies/subassemblies. Figure 29 is a global sequence visualization of the optic lens product.

(2) Linear comparison graphs

The linear comparison graph is intended to allow the user to make quick comparisons of assemblies based on the assembling steps, the number of reorientations, the assembly directions, and the limited precedence information. The linear comparison graph is shown with a three-axis format. There is a single axis for the x, y and z assembly directions. Since a part may move in a positive or negative direction on any of these axes this information is included in the label attached to the assembly step. Each assembly step is represented using a three dimensional CAD drawing that is placed on the axis corresponding to the direction that the part must move in order to be assembled. The graphical representation shows the assembly progressively growing from a single base part to a completed assembly at the end of a sequence. Reorientation of the product is easily recognized as either a change in

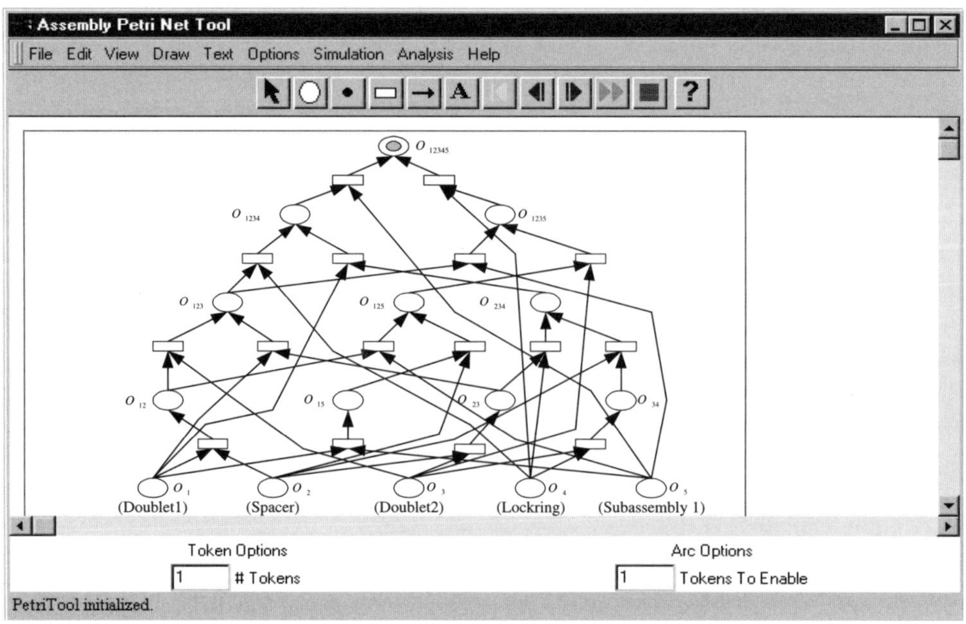

Fig. 29. Assembly sequence Petri net visualization.

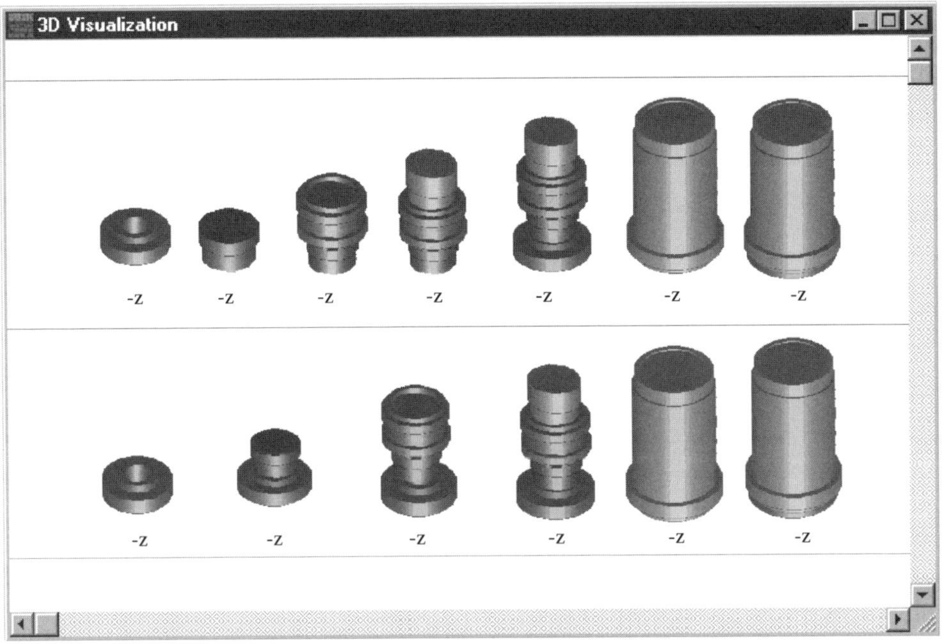

Fig. 30. Linear comparison graph of optic lens assembly.

the axis from x or y to z or it is encoded as a + or − sign in the label. It is also easy to compare the differences in assembly steps between the sequences by lining up the axes. The two assemblies may be scrolled simultaneously or individually in each window. Figure 30 shows the linear comparison graph for the optic lens, in which the assembly part sequence is the same, $O_4 \rightarrow O_3 \rightarrow O_2 \rightarrow O_1 \rightarrow O_5$, i.e. Lockring → Doublet2 → Spacer → Doublet1 → Subassembly1, but with different assembling steps.

(3) Snapshot graphs

The snapshot graphs provide the user with detailed part information and explicit precedence information and they are intended for detailed evaluations of particular sequences. The snapshot display integrates pictorial information into an arc and node diagrams of Petri net above. The resulting diagram displays a series of snapshots of the assembly actions in the appropriate order. The arc and node diagrams are appropriate for communicating the chronological order of the assembly. Abstract diagrams communicate abstract ideas, such as time, cost, assembly operation difficulty, etc. more efficiently than concrete pictures. The display allows engineers to examine the process in detail while maintaining a global view of the whole assembly. When the viewer selects a node in the diagram it will expand to show that node in detail while maintaining the node within the global view by shrinking the other nodes around it. This format allows the operator to adjust the display to an appropriate level of abstraction for the task at hand. The snapshot display supports the design process in several ways. First, it provides immediate images of the various assembly actions to enable engineers to have a better view of the process than they could visualize in their mind. Second, it supports a breadth-first approach by allowing the user to view a large portion of the sequence at one time without considering the details. Finally, it helps the user simulate the assembly process by providing a series of assembly act pictures. Figure 31 gives a snapshot graph of the optic lens assembly.

5.2.8. *Assembly simulation and animation module*

To verify and ascertain the effectiveness of the various models, the simulation of the assembly was carried out. With this simulation function, the users can observe the progress in the execution of the assembly Petri net, and they may also simulate the assembly system through virtual prototyping and virtual planning. Assembly simulation is closely related to the computer aided design and modeling system, the assembly planning system, and the assembly system design system.

The simulation of movement or property changes is of increasing importance for both 2D and 3D geometric applications. Designers will obtain the following benefits using animated machine modeling by[71]:

(1) Checking possible interference among moving parts in a modeling process;
(2) Simulating real working environments to improve the reliability of the design; and

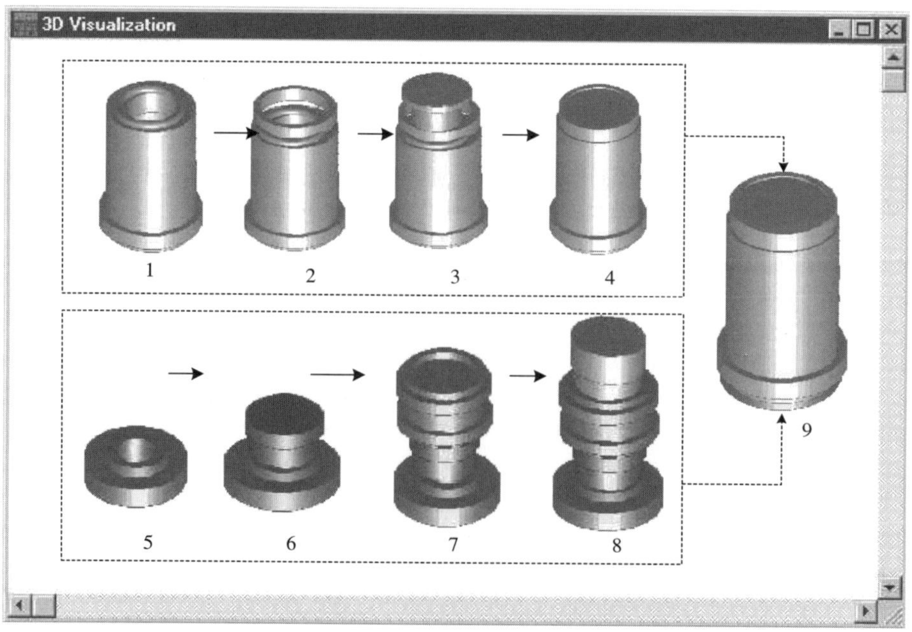

Fig. 31. Snapshot graph of optic lens assembly.

(3) Facilitating communications between designers, process planners and production managers particularly at an early design stage.

To visualize the assembly process, an interactive animation has been created for the proposed assembly plan. This animation includes the visualization of the tools (e.g. robot hand) and assembly. A complete and high fidelity visualization environment can be used not only by the design and process engineers while editing and simulating an assembly plan, but also by operators on the shop-floor to learn the assembly process.[71] The idea is to store the visualization information in a compact assembly-movie-format so that it can be viewed interactively.

The shop-floor visualization environment provides the assembly operator with a random-access 3D interactive animation of the complete assembly process.[71] In a compact format, the animation viewer stores the state of each component or tool as a function of time. This allows us to regenerate the state of the complete system (assembly workplace, tools, and the assembly components themselves) almost instantaneously, resulting in a random-access animation. The operator can jump forward or backward in time to a particular assembly operation of interest. To investigate the details of this operation, the operator can adjust the speed of the simulation continuously, pause it, and he can even move backwards in time. Furthermore, at any time during the animation, the operator can change the camera's viewpoint or zoom in to have a better study the details of the operation.

6. Conclusions

This chapter provided a knowledge-based intelligent framework for assembly oriented design and planning, which integrates the design, planning, and analysis processes of the electro-mechanical assemblies. The information and knowledge about a product and its assembly, e.g. assembly constraints, solid model and CAD data base, heuristic rules, etc., are represented by the integrated object model with a hierarchy of "place-transition" net structure, geometry, and feature models. Based on the assembly features and predicates, the object oriented, knowledge-based assembly modeling and reasoning can enhance the interactivity and integrity of assembly representation for assembly sequence planning. Integrated with the additive aggregation model of assembly operation's difficulty via the concept of the fuzzy sets theory, it can be used for assemblability and assembly sequence analysis and evaluation.

All feasible assembly sequences can be generated by reasoning and decomposing the leveled feasible subassemblies, and represented through Petri net. The use of Petri net models as a graphical language can specify and simulate the assembly process interactively and reduce computational burden. The decomposition of assembly plans based on Petri net representations obtains the lower-level assembly plans and the inherited properties such as liveness, safeness, and reversibility remains. The assembly sequence searching in Petri nets is mapped into a reachability problem.

The sequence of transitions, or the sequence of system states, i.e. the feasible components or operations sequences can be directly found from the reachability tree.

Qualitative strategic constraints are then used to evaluate the feasible assembly sequences. In order to obtain a good assembly sequence, some quantitative criteria such as assembly time, cost, operation difficulty, and part priority index are applied to select the optimal assembly sequence. Based on predefined time analysis and assemblability analysis, estimates are made on the assembly time, cost and the assembly operation's difficulty score of the product when each of these sequences is used. The optimal solution to the reachability problems due to the mapping and/or searching process is accomplished based on knowledge-based reasoning. Compared with the ordinary Petri net, the representation size of knowledge-based Petri net of the assembly sequence is the same, but the search space is much smaller due to many constraints and heuristic rules. Mostly the sequence obtained in knowledge-based Petri net model is optimal, and it is more economical this way.

The developed integrated knowledge-based assembly planning system (IKAPS) can achieve the integration of generation, visualization, evaluation and selection of assembly sequences in an integrated and interactive environment that helps the product engineer or the manufacturing engineer to carry out assembly planning in manufacturing systems. The application results show that the representational model for product assembly and assembly process are effective. The integrated method and knowledge-based system allow the design and analysis of the whole assembly process of a product, as well as the estimation of its assembly time and cost available in the early stages of product development. Future work is required on the incorporation of a function-based (function-behavior-structure) model into the integrated assembly model. Research is also needed on the further refinement of knowledge base in IKAPS with learning ability.

Disclaimer

The bulk of the work reported here by the author was conducted during his tenure at Nanyang Technological University and Singapore Institute of Manufacturing Technology, Singapore. Commercial equipment and software, many of which are either registered or trademarked, are identified in order to adequately specify certain procedures. In no case does such identification imply recommendation or endorsement by the National Institute of Standards and Technology, nor does it imply that the materials or equipment identified are necessarily the best available for the purpose.

References

1. G. Boothroyd and P. Dewhurst, *Product Design for Assembly*, (Boothroyd Dewhurst Inc., 1989).
2. S. S. Lim, I. B. H, Lee, L. E. N. Lim and B. K. A. Ngoi, Computer-aided concurrent design of product and assembly processes: a literature review, *J. Design and Manufacturing* **5** (1995) 67–88.

3. J. L. Nevins, D. E. Whitney, *Concurrent Design of Products and Processes*, (McGraw-Hill, 1989).
4. A. C. Lin, Automated Assembly Planning for Three-dimensional Mechanical Products, PhD Thesis, Purdue University, 1990.
5. H. J. Bullinger and M. Ritcher, Integrated design and assembly planning, *Computer Integrated Manufacturing Systems* **4**, 4 (1991) 239–247.
6. A. Delchambre, *Computer-aided Assembly Planning* (Chapman and Hall, 1992).
7. L. S. Homem de Mello and A. C. Sanderson, A correct and complete algorithm for the generation of mechanical assembly sequences, *IEEE Trans. Robotics and Automation* **7**, 2 (1991) 228–240.
8. L. S. Homem de Mello and A. C. Sanderson, Representation of mechanical assembly sequences, *IEEE Trans. Robotics and Automation* **7**, 2 (1991) 211–227.
9. T. L. De Fazio and D. E. Whitney, A prototype of feature-based design for assembly, *Proc. 16th Design Automation Conference (ASME), Advances in Design Automation*, Chicago, USA **32**, 1 (1990) 9–16.
10. K. Lee and D. C. Gossard, A hierarchical data structure for representing assemblies: Part 1, *Computer-aided Design* **17**, 1 (1985) 15–19.
11. J. D. Wolter, *On the Automatic Generation of Plans for Mechanical Assembly*, PhD Thesis, The University of Michigan, 1988.
12. L. S. Homem de Mello, *Task Sequence Planning for Robotic Assembly*, PhD Thesis, Carnegie Mellon University, 1989.
13. Y. F. Huang and C. S. G. Lee, Precedence knowledge in feature mating mating operation assembly planning, *IEEE Int. Conf. Robotics and Automation*, Scottsdale, Arizona (1989) 216–221.
14. U. Roy and C. R. Liu, Establishment of functional relationships between product components in assembly database, *Computer-aided Design* **20**, 10 (1989) 570–580.
15. C. J. M. Heenskerk and C. A. Van Luttervelt, The use of heuristics in assembly sequence planning, *Annals of the CIRP* **38**, 1 (1989) 37–40.
16. L. I. Liebermann and M. A. Wesley, AUTOPASS: An automatic programming system for computer controlled mechanical assembly, *IBM J. Research and Development* **21**, 4 (1977) 321–333.
17. Y. Liu and R. J. Popplestone, Planning for assembly from solid models, *IEEE Int. Conf. Robotics and Automation*, Scottsdale, Arizona (1989) 222–227.
18. J. Nieminen, J. Kanerva and M. Mantyla, Feature-based design of joints, *Advanced Geometric Modeling for Engineering Applications*, eds. F. L. Krause and H. Jansen, Berlin, Germany, (1989).
19. M. V. Rimscha, Feature modeling and assembly modeling—a unified approach, *Advanced Geometric Modeling for Engineering Applications*, eds. F. L. Krause and H. Jansen, Berlin, Germany, (1989).
20. R. E. Jones, T. L. Calton and R. R. Peters, Automated assembly and fixture planning at Sandia National Laboratories, *Assembly Automation* **17**, 3 (1997).
21. S. G. Kaufman, R. H. Wilson, R. E. Jones, T. L. Calton and A. L. Ames, The Archimedes 2 mechanical assembly planning system, *IEEE Int. Conf. Robotics and Automation* (1996) 3361–3368.
22. S. Lind and J. J. Gallimore, Generating assembly sequence diagrams using rules or previous diagrams, Wright State University Internal Report (Report 3), Department of Biomedical and Human Factors Engineering, Dayton, OH 45435, 1995.
23. R. H. Wilson and J. F. Rit, Maintaining geometric dependencies in an assembly planning, *Computer-Aided Mechanical Assembly Planning*, eds. L. S. Homem de Mello and S. Lee (Kluwer Academic Publishers, USA, 1991) 217–241.

24. R. H. Wilson, *On Geometric Assembly Planning*, PhD Thesis, Department of Computer Science, Stanford University, 1992.
25. T. Lozano-P'erez, *The Design of a Mechanical Assembly System*, Technical Report AI-TR 397 (AI Laboratory, MIT Press, 1983).
26. R. H. Taylor, Synthesis of Manipulator Control Programs from Task-level Specifications, PhD Thesis, Department of Computer Science, Stanford University, 1976.
27. J. C. Latombe, *Robot Motion Planning* (Kluwer Academic Publishers, Boston, MA, 1991).
28. A. Bourjault, *Contribution a une Approache Methodologique de L'assemblage Automatise: Elaboration Automatique des Sequences Operatioires*, These d'Etat, Universite de Franche-Comte, Besancon, France, 1984.
29. T. L. De Faxio and D. E. Whitney, Simplified generation of all mechanical assembly sequences, *IEEE J. Robotics and Automation* **RA-3**, 6 (1987) 640–658.
30. D. F. Baldwin, *Algorithmic Methods and Software Tools for the Generation of Mechanical Assembly Sequences*, Master's Thesis (MIT, MA, USA, 1990).
31. R. L. Hoffman, A Common sense approach to assembly sequence planning, *Computer-Aided Mechanical Assembly Planning*, eds. L. S. Homem de Mello and S. Lee, Boston, MA (Kluwer Academic Publishers, 1991) 289–314.
32. S. Lee and Y. G. Shin, Assembly planning based on geometric reasoning, *Computation and Graphics* **14**, 2 (1990) 237–250.
33. B. Romney, C. Godard, M. Goldwasser and G. Ramkumar, An efficient system for geometric assembly sequence generation and evaluation, *Proc. 1995 ASME Int. Computers in Engineering Conf.* (1995) 699-712.
34. G. Boothroyd and L. Alting, Design for assembly and disassembly, *Annals of the CIRP* **41**, 2 (1992) 625–636.
35. M. J. Jakiela and P. Papalambros, Design and implementation of a prototype intelligent CAD system, *ASME J. Mechanisms, Transmission, and Automation in Design* **111**, 2 (1989).
36. W. Hsu, C. S. G. Lee and S. F. Su, Feedback approach to design for assembly by evaluation of assembly plan. *Computer-aided Design* **25**, 7 (1993) 395–410.
37. R. H. Sturges and M. I. Kilani, Towards an integrated design for an assembly evaluation and reasoning system, *Computer-aided Design* **24**, 2 (1992) 67–79.
38. T. H. Liu, *An Object-oriented Assembly Applications Methodology for PDES/STEP-based Mechanical Systems*, PhD Thesis, The University of Iowa, USA, 1992.
39. T. H. Liu and G. W. Fischer, Assembly evaluation method for PDES/STEP-based mechanical systems, *J. Design and Manufacturing* **4** (1994) 1–19.
40. L. M. Rosario, *Automatic Geometric Part Features Calculation for Design for Assembly Analysis*, PhD Thesis, University of Rhode Island, USA, 1988.
41. D. Ben-Arieh and B. Kramer, Computer-aided process planning for assembly: generation of assembly operation sequences, *Int. J. Production Research* **32**, 3 (1994) 643–656.
42. P. Gu and X. Yan, CAD-directed automatic assembly sequence planning, *Int. J. Production Research* (1996) 3069–3100.
43. S. Lee and Y. G. Shin, Assembly co-planner: cooperative assembly planner based on subassembly extraction, *Computer-Aided Mechanical Assembly Planning*, eds. L. S. Homem de Mello and S. Lee (Kluwer Academic Publishers, USA, 1991) 315–339.
44. M. Shpitalni, G. Elber and E. Lenze, Automatic assembly of three dimensional structures via connectivity graphs, *Annals of the CIRP* **38**, 1 (1989) 25–28.

45. S. Lee and Y. G. Shin, Automatic construction of assembly partial order graphs, *Int. Conf. Computer Integrated Manufacturing*, Rensselaer Polytechnic Institute (1988) 383–392.
46. H. J. Bullinger and E. D. Ammer, Computer aided depicting of precedence diagrams—a step towards efficient planning in assembly, *Computing and Industrial Engineering* **18**, 3/4 (1984) 165–169.
47. J. D. Wolter, On the automatic generation of assembly plans, *Computer-Aided Mechanical Assembly Planning*, eds. L. S. Homem de Mello and S. Lee (Kluwer Academic Publishers, USA, 1991) 263–288.
48. De Floriani and G. Nagy, A graph model for face-to-face assembly, *Proc. IEEE Int. Conf. Robotics and Automation* **1** (1989) 75–78.
49. W. X. Zhang, Representation of assembly and automatic robot planning by Petri net, *IEEE Trans. System, Man, Cybernetics* **29**, 2 (1989) 418–422.
50. L. Laperriere and H. A. Eimaraghy, Planning of product assembly and disassembly, *Annals of the CIRP* **41**, 1 (1992) 5–9.
51. L. Laperriere and H. A. Eimaraghy, Assembly sequences planning for simultaneous engineering applications, *Int. J. Advanced Manufacturing Technol.* **9** (1994) 231–244.
52. R. B. Gottipolu and K. Ghosh, An integrated approach to the generation of assembly sequences, *Int. J. Computer Appl. Technol.* 1 (1996) 135–138.
53. S. Kanai, H. Takahashi and H. Makino, ASPEN: computer-aided assembly sequence planning and evaluation system based on predetermined time standard, *Annals of the CIRP* **45**, 1 (1996).
54. B. J. Jeong, A Computer-Aided Evaluation System for Assembly Sequences in Flexible Assembly Systems, PhD Thesis, The Pennsylvania State University, USA, 1993.
55. J. L. Peterson, *Petri Nets Theory and the Modeling of Systems* (Englewood Cliffs, NJ: Prentice-Hall, 1981).
56. H. K. Tonshoff, U. Beckendorff and M. Schaele, Some approaches to represent the interdependence of process planning and process control, *Proc. 19th CIRP Int. Seminar Manufacturing Syst.* State College, Pennsylvania (1988) 165–170.
57. D. Ben-Arieh, A methodology for analysis of assembly operations' difficulty, *Int. J. Production Res.* **32**, 8 (1994) 1879–1895.
58. D. S. Hong and H. S. Cho, A neural-network-based computational scheme for generating optimised robotic assembly sequences, *Eng. Appl. Artificial Intelligence* **8**, 2 (1995) 129–145.
59. G. Hird, K. G. Swift, R. Bassler, U. A. Seidel and M. Richter, Possibilities for integrated design and assembly planning, *Proc. 9th Int. Conf. Assembly Automation*, London, UK, ed. A. Pugh, Developments in Assembly Automation, Springer Verlag, IFS Publications, Bedford, UK (1988) 155–166.
60. J. M. Henrioud and A. Bourjault, LEGA: a computer-aided generator of assembly plans, *Computer-Aided Mechanical Assembly Planning*, eds. L. S. Homem de Mello and S. Lee (Kluwer Academic Publishers, USA, 1991) 191–215.
61. M. Santochi and G. Dini, Computer-aided planning of assembly operations: the selection of assembly sequences, *Robotics and Computer-Integrated Manufacturing* **9**, 6 (1992) 439–446.
62. C. L. P. Chen and Q. W. Yan, Design of a case associative assembly planning system, *Intelligent Eng. Syst. Through Artificial Neural Networks*, eds. C. H. Dagli, S. R. Y. Kumara and Y. C. Shin (New York: ASME, 1991) 757–762.
63. C. L. P. Chen, Intelligent assembly system for sequence generation and task planning, *Intelligent Syst. Design and Manufacturing*, eds. C. H. Dagli and A. Kusiak (1994) 319–357.

64. D. F. Baldwin, *et al.*, An integrated computer aid for generation and evaluation assembly sequences for mechanical products, *IEEE Trans. Robotics and Automation* **7**, 1 (1991) 78–94.
65. R. Karinithi and D. N. Nau, Geometric reasoning using a feature algebra, *Artificial Intelligence Applications in Manufacturing*, eds. A. Famili, D. S. Nau and S. H. Kim (1989) 41–60.
66. A. C. Lin and T. C. Chang, An integrated approach to automated assembly planning for three dimensional mechanical products, *Int. J. Production Res.* **31**, 5 (1993) 1201–1226.
67. J. K. Gui, *Methodology for Modelling Complete Product Assemblies*, PhD Dissertation, Helsinki University of Technology, 1993.
68. S. Gottschlich, C. Ramos and D. Lyons, Assembly and task planning: a taxonomy and annotated bibliography, *IEEE Assembly and Task Planning Technical Committee* (1993).
69. T. H. Cao and A. C. Sanderson, Intelligent task planning using fuzzy Petri nets, *World Scientific, Intelligent Control and Intelligent Automation* **3** (1996).
70. J. C. Giarratano, CLIPS Reference Manual: Volume I, II, III, Basic Programming Guide, CLIPS version 6.0. Software Technology Branch, Lyndon B. Johnson Space Center, Houston, TX, 1993.
71. S. K. Gupta, C. J. J. Paredis and R. Sinha, An intelligent environment for simulating mechanical assembly operations, *Proc. DETC'98 1998 ASME Design Eng. Tech. Conf.* Atlanta, Georgia, USA, 1998.
72. M. H. Overmars and P. Svestka, A probabilistic learning approach to motion planning, *Proc. First Workshop on Algorithmic Foundation of Robotics* (A. K. Peeters, Boston, MA, 1994).
73. J. P. Thomas, A Petri net framework for representing mechanical assembly sequences, *IEEE/RJS Int. Conf. Intelligent Robots and Systems*, Raleigh, NC (1992) 2116–2121.
74. J. P. Thomas, Constructing assembly plans, *IEEE Int. Conf. Robotics and Automation*, Atlanta, GA (1993) 515–520.
75. R. S. Groppetti and S. N. Antonio, On the application of coloured Petri nets to computer aided assembly planning, *IEEE Symp. Emerging Technologies and Factory Automation* (1994) 381–387.
76. S. Caselli and F. Zanichelli, On assembly sequence planning using Petri nets, *Proc. IEEE Int. Symp. Assembly and Task Planning* (1995) 239–244.
77. T. Cao and A. C. Sanderson, Task sequence planning in a robot workcell using AND/OR nets, *IEEE Int. Symp. Intelligent Control*, Arlington, VA (1991) 239–244.
78. T. Suzuki, T. Kanehara, A. Inaba and S. Okuma, On algbraic and graph structural properties of assembly Petri nets, *IEEE Int. Conf. Robotics and Automation*, GA (1993) 507–514.
79. T. Suzuki, T. Kanehara, A. Inaba and S. Okuma, On algebraic and graph structural properties of assembly Petri nets-searching by linear programming, *IEEE/RSJ Int. Conf. Intelligent Robots and Systems*, Raleigh, NC (1992) 2116–2121.
80. T. Murata, Petri nets: properties, analysis and applications, *Proc. IEEE* **77**, 4 (1989) 541–580.
81. J. Martinez and M. Silva, A simple and fast algorithm to obtain all in-variants of a generalized Petri net, *2nd European Workshop on Application and Theory of Petri Nets* (Springer-Verlag, 1981).
82. C. V. Ramamoorthy and G. S. Ho, Performance evaluation of asynchronous concurrent systems using Petri nets, *IEEE Trans. Software Engineering* **SE-6**, 5 (1980) 440–449.

83. N. Deo, *Graph Theory with Applications to Engineering and Computer Science* (Englewood Cliffs, N.J., Prentice-Hall, 1974).
84. M. Futtersack and J. M. Labat, Constraint programming within CLIPS, *http://www.droit.univparis5.fr/futtersack/english/research/CCP/index.html*, 1997.
85. X. F. Zha, A new approach to generation of ruled surfaces and its applications in engineering, *Int. J. Advanced Manufacturing Technol.*, UK **13**, 3 (1997).
86. X. F. Zha and W. M. Jin, Screw method for robot path generation, *J. Southeast University, China* (English Version) **18**, 1 (1992) 116–123.
87. X. F. Zha, H. L. Du, X. F. Wang and W. M. Jin, Using the theory of dual transformation to plan and analyze the robot trajectory motion, *J. Robotics* **15**, 1 (1993) 16–21.
88. C. K. Choi, X. F. Zha, T. L. Ng and W. S. Lau, On the automatic generation of assembly sequence, *Int. J. Production Res.* **3** (1998) 617–633.
89. X. F. Zha, *Knowledge Intensive Methodology for Intelligent Design and Planning of Assemblies*, PhD Thesis, Nanyang Technological University, Singapore, 1999.
90. X. F. Zha, S. Y. E. Lim and S. C. Fok, Integrated intelligent design and assembly planning: a survey, *Int. J. Advanced Manufacturing Technol.* **14**, 9 (1998) 664–685.
91. X. F. Zha, S. Y. E. Lim and S. C. Fok, Integrated knowledge-based assembly sequence planning, *Int. J. Advanced Manufacturing Technol.* **14**, 1 (1998) 50–64.
92. X. F. Zha, S. Y. E. Lim and S. C. Fok, Integrated knowledge-based Petri net intelligent flexible assembly planning, *J. Intelligent Manufacturing* **9**, 3 (1998) 235–253.
93. X. F. Zha, S. Y. E. Lim and S. C. Fok, Integrated knowledge-based approach and system for product design for assembly, *Int. J. Computer Integrated Manufacturing* **12**, 3 (1998) 211–237.
94. X. F. Zha, S. Y. E. Lim and S. C. Fok, Concurrent integrated design and assembly planning, *Proc. 4th Int. Conf. Automation, Robotics, and Computer Vision* (ICARCV'96), Singapore, 1996.
95. X. F. Zha, S. Y. E. Lim and S. C. Fok, Development of expert system for concurrent product design and planning for assembly, *Int. J. Advanced Manufacturing Technol.* **2**, 5 (1999) 153–162.
96. http://www.cs.wright.edu/research/caap/
97. R. E. Douglas, *SNEAKERS: A Concurrent Engineering Demonstration System*, MS Thesis, Worcester Polytechnic Institute, 1993.

CHAPTER 2

NEURAL NETWORKS TECHNIQUES FOR THE OPTICAL INSPECTION OF MACHINED PARTS

NICOLA GUGLIELMI

Dipartimento di Matematica Pura e Applicata,
Università dell'Aquila,
via Vetoio (Coppito),
67100 L'Aquila, Italy
E-mail: guglielm@univaq.it

ROBERTO GUERRIERI and GIORGIO BACCARANI

D.E.I.S. Università di Bologna,
Viale Risorgimento 2,
40136 Bologna, Italy

This paper presents a survey on neural-network techniques for the optical inspection of machined parts. In the first part we deal with some popular neural-based methodologies for the quality control of machined surfaces, and then we focus on some general problems linked to this type of applications, such as the relative scarcity of examples, commensurate to the high dimensionality of the input space, and the exploitation of *a priori* information. In the second part, we present two complete applications of neural networks to deal with specific industrial quality-control problems, and then we perform a full classification starting from the raw image all the way down to the final detection flag which indicates whether there is a failure in the original image. More specifically, the first application uses only highly constrained neural networks embedding a large amount of *a priori* information, which is available in terms of spatial and structural invariances. This network has been successfully trained in spite of a significant scarcity of training samples. The second application, instead, is a typical example of the use of adaptive techniques in constrained hierarchic frameworks.

Keywords: Numerical methods for image processing; neural networks; adaptive numerical filtering; constrained optimization; optical inspection; pattern analysis.

1. Introduction

The visual inspection of machined parts is a powerful technique for quality control, as it allows for an unexpensive and non-destructive investigation of a machined surface to reveal the presence of structural defects which need to be trapped in the early stages of the manufacturing process.

An important problem related with this technique is linked to the significant cost of developing the software to interpret the images. Most of the "machine vision systems" for the automated industrial inspection are custom-designed. Their rather inflexible structure prevents them to be adapted to different tasks, and sometimes even preventing them from operating on the same application under a different environment.[1] Furthermore, due to the high specificity of the required solutions, every detection system has a low diffusion volume, which increases the software cost that is divided among a few customers.

A second consideration concerns the high reliability required by the system under real-time operating conditions. This system requirement will in turn demand the use of high quality software. Since a large part of the code has to be finely tuned to the specific task at hand, the software development cannot cover a large spectrum of applications.

A third aspect is that, although neural networks are particularly well suited to parallel machines, most of their pattern-recognition applications are only able to process small-scale images. Most large-scale image patterns are recognized by extracting special features. This approach is limited because it is difficult to find general methods to extract meaningful characteristics. In general, NN-application in image identification is hard to be implemented in large scale data because the training period is directly related to the number of pixels used to parametrize the image.[2] This situation is dramatically worsened when the number of "classification examples" is too small, and thus insufficient for a correct training of the recognition system.

This paper is organized as follows. In Sec. 2 we describe the recognition problem and some of the research lines developed in the literature, with main reference to the adaptive systems. In Sec. 3 we discuss how the quality control task can be formulated in terms of a diagnosis problem. Several papers which have appeared recently in literature, are quoted to provide a wide spectrum of new ideas. In Sec. 4 we review, in the light of the given discussion, a real-life application recently considered by the authors of Ref. 3. Then, in Sec. 5, we describe a second application pertaining to the quality analysis of printed circuits. Finally, some conclusion are drawn in Sec. 6.

2. Adaptive Recognition Systems for Optical Inspection of Machined Parts

Following a consolidated line, we can summarize the main features of the industrial optical inspection of machined parts as follows:

(i) The material flux is continuous;
(ii) The flow rate is relatively high (typically 0.5–5 meters per second);
(iii) A high spatial resolution is required (some hundreds of pixels per decimeter);
(iv) Defects typically cover small fractions of the total surface area;
(v) The same defect exhibits different patterns.

Knowledge-based machine-vision systems have been applied to a wide spectrum of different inspection tasks, while other non-customized solutions aim for a reduced flexibility but with increased efficiency and speed.

The use of neural networks provides better flexibility, allowing us to adapt the system to a number of similar applications with ease through simple parameter changes. This is a good compromise between highly specific systems and general-purpose ones. Flexibility in terms of different applications may be problematic instead, since it generally prevents the required efficiency. For a wide survey on machine-vision systems for industrial inspection we refer the reader to Ref. 4.

2.1. *Object analysis*

The goal of computer-aided image analysis is to automatically recognize specific objects in the image. To do this, the computer has to learn about the patterns in the image. According to Winston,[5] there are several learning strategies: learning by being programmed, learning by being told, learning by observing samples and learning by discovery.

The first strategy is based on experts in the field who developed a set of routines, finely tuned to the specific goals. Whenever there are new kinds of images to be processed, new algorithms are implemented and added to the program. By coupling the expert systems to the sophisticated image processing libraries, the second strategy makes it easier to guide the user in selecting suitable algorithms to deal with specific classification tasks. As an example, we quote the recognition system IMARS (Interactive Modeling and Automatic Recognition System) as described by Tomita[6] and the system developed in Ref. 7, which was applied in the investigation of machined-parts. The third kind of approach has been widely exploited in the last few years and it is based on the powerful learning capabilities of neural networks.[8,9] Finally, the fourth approach is presently being investigated by numerous researchers and some very interesting problems concerning the knowledge and inference bases have been found.

2.2. *A survey of neural-network techniques for quality control of machined surfaces*

The application of 2D image techniques to the visual inspection of flawed surfaces has been widely considered. These methods are effective in enhancing the contrast between the flaws (such as cracks) and the background but they do not address, by themselves, the identification and the classification of the flaws. Hence they are preprocessing tools, leading to enhanced images for a further identification, which is often based on neural-network systems.

Standard neural approach

A CAD-based inspection methodology has been proposed by Newman and Jain[10] in order to inspect planar, cylindrical and spherical surfaces. This technique is able to detect various kinds of defects and is based on a first segmentation and a subsequent clustering of regions having similar geometrical and statistical properties. However, several problems were encountered and the construction of the CAD models was very time consuming.

To improve the efficiency of this kind of systems, classical solutions based on intensity histograms have been proposed, for example, by Bradley and Kurada.[11] Their technique makes use of some low-level software, which separates the aforementioned image segmentation into distinct surface patches and then performs a histogramming on each patch based on the gray-level intensity. Since the histograms are mapped into one dimensional vectors, they are very well suited to a neural-network classification.

More specifically, Bradley and Kurada proposed a back-propagation network having three layers of sigmoid transfer functions; this solution was mostly selected from experience.

Skeleton and neural-network based approach

Skeletonization is useful for reducing the number of pixels required to represent images. The skeleton function is particularly sensitive to sharp changes in the intensity of the image, like those associated with the presence of cracks. Due to this reason, utilizing the skeleton function leads to robust neural networks requiring limited training and execution time (see, for example Refs. 2 and 12). This is a very promising image-processing technique because the skeleton of an object consists of all the information related to its geometric properties in a reduced form. A possible problem arises since skeletonization is not rotationally invariant. This concept of rotational invariance has been discussed by several authors and is essentially due to the fact that the *structuring elements*, based on the pixels, are not disk shaped. However, there are techniques which make skeletonization substantially invariant with respect to the possible rotations of the objects.

This approach typically consists of two steps. The first stage performs a nearly rotation-invariant skeletonization (see e.g. Refs. 13 and 14). The second stage provides a sequential coding of the obtained objects, aiming to provide a *normalized* parametrization of the object, in order to get a fixed number of inputs to a supervised neural network. An approach which makes use of a skeletonization step is also discussed in Ref. 7.

Neural associative modules

König *et al.*[15] discuss the application of neural and associative modules in the context of a hybrid dynamic system for visual object inspection in industrial quality control. Generally, automatic quality control (AQC) systems consist of an image acquisition and a processing front end. Based on a set of computed features, the classification of the different regions is performed. Rule-based systems are often effective in these applications; a set of rules can in fact be provided by the experts in such a way that the classification region can be inferred from the application of these rules. It is clear that these rules cannot cover all situations that might arise during the real inspection. Therefore, the definition of a fixed set of rules is not sufficient; moreover, in many cases such rules are not completely known. Neural associative memories provide those generalization properties which allow the recognition of patterns that are not precisely described by a predefined set of rules. The combination of rule-based systems and neural associative memories is able to exploit the available knowledge and to extract part of the *implicit* knowledge of the problem.

König *et al.* studied the implementation of hybrid systems with self-organizing properties, incorporating image processing modules, feature extraction modules and characterized by a mixed knowledge base, a neural-network and a neural associative memory topology.

As a further application, Poechmueller *et al.*[16] present an approach to the classification of solder joint images by means of a neuron-like associative memory. It is interesting that all the algorithms proposed could be easily implemented with the digital VLSI technology, thus providing a compact and fast classifier.

Image correlation neural-network techniques

Taking a whole-image information into account for visual inspection is quite difficult because of the essential lack of fault tolerance. The neural-network image-correlation technique tries to overcome this difficulty by using artificial neural networks without the usual interconnected cellular structure but "equipped with the automatic feature-extraction functionality, which can synthesize the relative image correlation, that is immuned to the absolute registration of a single frame".[17] The key idea in developing a self-referenced filtering among a sequence of frames is to be able to extract the correct statistical and geometric features of the scene, instead of processing the individual variable frame.

A sequence of frames requires the suitable tracking of the objects in the image domain. Such an operation provides the self-reference matched filter, which minimizes the digitization and registration error and noise (see Ref. 18).

In particular, we refer the reader to the paper by Villalobos and Gruber[19] who used this technique to specifically design a system for neural-network based inspection of machined surfaces. This technique is a useful method for on-line inspection of quality-production control.

3. The Classification Problem: A Theoretical Framework

3.1. *The recognition system*

There are two main strategies for "texture analysis": statistical analysis and structural analysis. The former is based on scalar measurements on the image, while the latter is based on structural descriptions. Both approaches can be implemented in adaptive ways by providing a learning phase. The processing flow of an adaptive statistical-based image classification is summarized by the following steps:

(i) **Image filtering**: some filters are applied to the image to enhance the patterns to be investigated. This phase must be finely tuned in order to eliminate most of the noise and the *apparent* defects.
(ii) **Image segmentation**: the system segments the image into a number of regions characterized by uniform gray-level intensities.
(iii) **Grouping**: the system decomposes the image into a number of connected regions. Sometimes complex shapes are decomposed into simpler ones.
(iv) **Statistical analysis**: the system computes a number of scalar properties of every input pattern by statistical texture analysis. These properties are used to provide a first classification of the image into the category which has more similar properties. If there are no doubts about *labelling* the image the analysis will stop immediately. Otherwise, if there is competition between several categories, the analysis has to advance further. The role of this first phase is for the pre-selection of candidate categories.
(v) **Pattern parametrization**: every pattern is parametrized and clustered into a features vector.
(vi) **Classification**: this phase consists of two moments:
    (a) **Learning phase**: the training patterns, which have been previously labelled, are used to train the network to adapt itself to the specific task.
    (b) **Recognition phase**: every input image is analysed to be classified into the proper category.

An example is illustrated in Fig. 1.

The processing flow of adaptive structural image classification is summarized, instead, by the following steps:

(i) **Choice of a general architecture**: in the first phase the number of layers and the number of processing units are chosen to process the input data. The design of the network should possess the capabilities to capture the essential features of the input patterns.
(ii) **Constraining the network**: in this phase, the *a priori* information is *structurally* included into the network as suitable constraints on the class of functions represented by the system.
(iii) **Learning phase**: in this phase the network is trained by a set of examples to determine the values of the parameters of the system. As opposed to the previous case, no parametrization of the processed patterns into classification vectors is done. The processed information flows directly through the network till the final layer, in the form of an output pattern.

The processing flow is illustrated in Fig. 2.

In Refs. 3 and 7 a comparison between these strategies is presented.

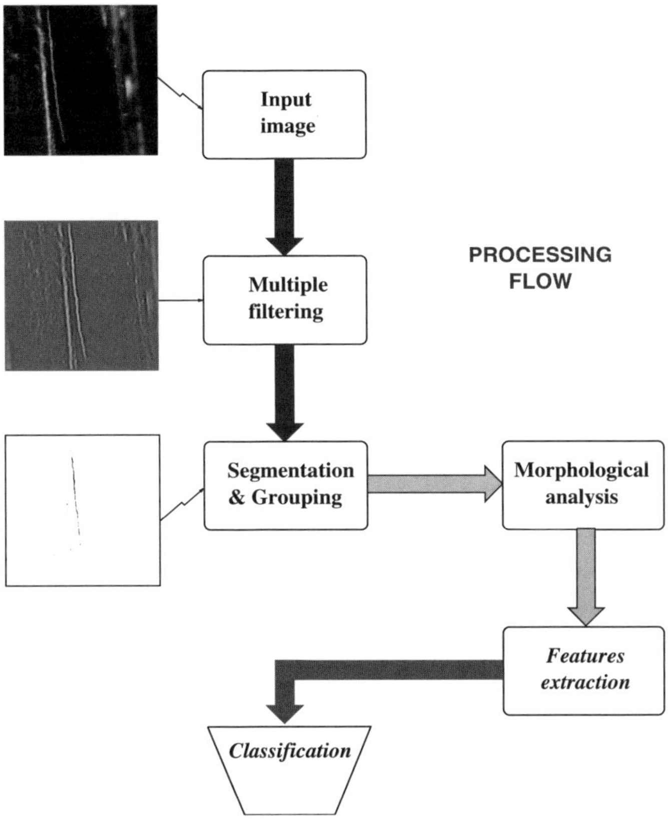

Fig. 1. An example of the first strategy.[3]

3.2. *The choice of neural networks*

The object recognition problem is a typical task where neural networks exhibit their powerful capabilities. In this paper, we focus our attention on the class of *supervised* neural networks, which can be trained to provide a classification which is known *a priori*. The typical structure of such a network is given by a number of layers. The input information, which is stored to form suitable patterns, is fed to the input layer, giving rise to a computational flow, and then terminating at the output layer, which provides the *classification* (still in the form of a suitable pattern). The training phase has the goal of *teaching* the network the relationships between the input and the output data.

At the end of this learning phase, it is essential to check that the NN has correctly *generalized* such relationships. This is done through several error computations, mainly based on L_2-measures. The Error Back Propagation learning algorithm (see, for example Refs. 20 and 21) is one of the most consolidated techniques used to train neural networks for many applications. This algorithm essentially implements

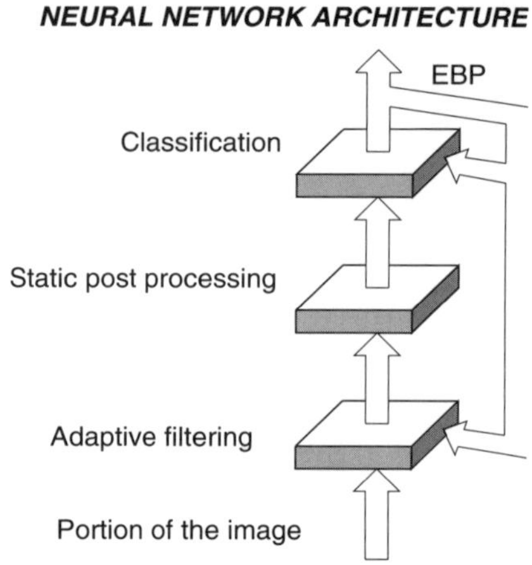

Fig. 2. The possible processing flow in the second strategy.

a gradient-based least square minimization of the error function, which is obtained by computing the distance between the outputs provided by the NN and the *desired* outputs.

3.3. *Neural networks as optimal estimators*

Feedforward neural networks, trained by error back propagation,[20] provide an example of nonlinear regression estimators.

The classification problem we consider can be formally described in the following way. Given an input vector $\mathbf{x}$ and an output vector $\mathbf{y}$, we consider the goal to predict $\mathbf{y}$ by $\mathbf{x}$. In the context considered here, it is not restrictive to assume $\mathbf{y}$ to be a scalar, and then identifying one of the $\mathcal{K}$ classes ($\mathcal{K}$ is an integer).

The learning problem consists of building a function $f(\mathbf{x})$, on the basis of the observed data $\{(\mathbf{x}_i, y_i)\}$, $i = 1, \ldots, N$ such that $f(\mathbf{x})$ approximates, according to some criteria, the desired answer y.

A well-known choice is to associate the problem with a function, say J, which is to be minimized. For example, in the *least square* error minimization, the choice is

$$J = \sum_{i=1}^{N} [y_i - f(\mathbf{x}_i)]^2 \,. \tag{1}$$

In the case of neural networks, f turns out to be parametrized by a certain number of so-called *synaptic weights*,

$$y = f(\mathbf{x};\, \mathbf{w}) \,,$$

where $\mathbf{w}$ is a vector of s real components, providing the free parameters of the system. Hence, the cardinality of $\mathbf{w}$ restricts the determination of the function f into a certain subclass of *possible* functions. By setting $\mathbb{N}_{\mathcal{K}} = \{1, \ldots, \mathcal{K}\}$ and by denoting the dimension of the input vector $\mathbf{x}$ with n and the corresponding input space with X^n, we have:

$$f:\ X^n \times \mathbb{R}^s \longrightarrow \mathbb{N}_{\mathcal{K}}\,.$$

In a probabilistic interpretation (see Refs. 9 and 22), the minimization of J triggers a regression estimation to compute the function of $\mathbf{x}$, giving the mean value of y conditioned on $\mathbf{x}$. The main motivation of this choice lies with the advantage that the regression $E[y|\mathbf{x}]$ yields the best estimator of y conditioned on $\mathbf{x}$, over the set of all functions, in the sense of least squares optimization.

As suggested by many authors (e.g. Ref. 22), we can interpret the classification process in the following sense. Given a sequence of *observations* $\{(\mathbf{x}_i, y_i)\}$, we estimate:

$$P(k|\mathbf{x}) = E[k|\mathbf{x}]\,,$$

where E represents the expected value with respect to the unknown probability distribution and $k \in \mathbb{N}_{\mathcal{K}}$ denotes the kth class. The classification is then obtained through a decision rule:

$$\mathbf{x} \in k \qquad \text{if} \qquad P(k|\mathbf{x}) > \theta\,,$$

with θ as a suitable probability threshold. This technique induces a separation within the space of parameters, which is the goal of the estimation procedure.

Of course, several approaches can be used on this technique. A first methodology, which is said to be *parametric*, assumes that the data belong to *a priori* known classes depending on the (possibly small) number of parameters (e.g. mixtures of Gaussians). Consequently it is natural to proceed to the Bayesian classification. A second methodology, which is said to be *nonparametric*, does not make instead any *a priori* assumptions.

A first important aspect to consider is consistency, that is the convergence of an estimator (f) to the function to be estimated ($E[y|\mathbf{x}]$) as $N \to \infty$. Widely used non-parametric algorithms demonstrating this property are the k-nearest neighbor algorithm (see, for example Refs. 23 and 24) and the projection pursuit technique.[25]

The main problem of these estimators is the necessity in having a progressively increasing number of examples available (according to the required accuracy), which is often impossible in industrial tasks. A second problem, linked to the previous one, is the slow convergence, and the consequent computational effort, which prevents real-time processing.

Due to these reasons, when the training set has a small fixed size, a parametric estimator easily overcomes a non-parametric one, even when the *true* probability function does not belong to the class parametrized by the estimator.

According to Geman *et al.*,[9] the problem exists in the estimation of $f(x)$, on the basis of a training set $\mathcal{D} = \{(x_1, y_1), \ldots, (x_n, y_n)\}$, with the aim of evaluating y for future observations of x. To clarify the dependence of f on the data, we shall denote by $f(x; \mathcal{D})$ the function to be estimated. Given an input x and a data set $\mathcal{D}$, a measure of the efficiency of f as a predictor of the output y is the mean squared error,

$$E[(y - f(x;\mathcal{D}))^2|x,\mathcal{D}]\,. \tag{2}$$

Then, expanding Eq. (2) yields (see Ref. 9)

$$E[(y - f(x;\mathcal{D}))^2|x,\mathcal{D}] = E[(y - E[y|x])^2|x,\mathcal{D}] + (f(x;\mathcal{D}) - E[y|x])^2\,.$$

The first term on the right hand side does not depend on either the data $\mathcal{D}$ or the estimator f. Consequently, the distance

$$(f(x;\mathcal{D}) - E[y|x])^2$$

seems to be a *natural measure* of the efficiency of f as an estimator of y.

Now let $E_{\mathcal{D}}[(f(x;\mathcal{D}) - E[y|x])^2]$ be the average over possible sets of data $\mathcal{D}$ (with N fixed). Then the *bias/variance* decomposition of Geman *et al.* is given by

$$\begin{aligned} E_{\mathcal{D}}[(f(x;\mathcal{D}) - E[y|x])^2|x,\mathcal{D}] &= (E_{\mathcal{D}}[f(x;\mathcal{D})] - E[y|x])^2 \\ &\quad + E_{\mathcal{D}}[(f(x;\mathcal{D}) - E_{\mathcal{D}}[f(x;\mathcal{D})])^2]\,, \end{aligned}$$

where the first addendum denotes the *bias*, while the second one stands for the *variance.*

If, on the average, $f(x;\mathcal{D})$ is different from $E[y|x]$, we say that $f(x;\mathcal{D})$ is a biased estimator. Viceversa, if the variance term prevails, the estimator may depend critically upon the data.

3.3.1. *Neural estimators*

This is substantially an intermediate choice, which tries to balance a rather general *a priori* knowledge (for example the regularity of the data) with the faithfulness of the observed data. The output of a feed-forward neural network is

$$y = f(x; w(\mathcal{D}))\,,$$

where $w(\mathcal{D})$ represents the *weights* of the system, to be specified on the basis of $\mathcal{D}$. The determination of the optimal weights is obtained through the approximate minimization of the sum of squared errors,

$$\sum_{i=1}^{N}(y_i - f(x_i; w(\mathcal{D})))^2\,.$$

If the number of weights is small, the class of spanned functions will turn out to be quite limited. Consequently the neural network is probably biased. On the contrary, if the number of weights is too high, which means an overparametrization via a large number of connections, the bias is reduced, but the risk of a significant contribution to the variance component arises.

When we consider problems characterized by a high dimensionality of the input space, the use of neural networks is often connected (in a natural way) to the use of a large number of interconnections and, consequently, weights. Hence, when applied to industrial classifications with a (relatively) small number of available examples, it is necessary to proceed to a progressive biasing of the network, based — whenever possible — on knowledge-based rules, and then reaching a tradeoff between biased and high variance solutions.

A possible *reasonable criterion*, which has the goal of balancing bias and variance, consists of the cross validation.[26] Given a training set $\mathcal{D} = \{(x_i, y_i)\}_{i=1}^{N}$, the general estimator is denoted by $f(x; N, \mathcal{D})$. The cross validation is based on the leave-one-out estimation principle. If $\mathcal{D}_N^i$ is the set of data $\mathcal{D}$ with a sample removed (x_i, y_i), and if the estimator $f(x_i; N-1, \mathcal{D}_N^i) \approx y_i$, then we have a result in favor of the estimator $f(x; N, \mathcal{D})$ since, in this case, we do not expect substantial differences between the two estimators.

3.4. *A correct use of Error Back Propagation*

An efficient use of the EBP algorithm depends on the correct choice of the number of layers and neural units and the suitable parametrization of the input-data. For example, the choice of an optimal number of internal layers has been investigated by many authors (see e.g. Cybenko[27]).

Furthermore, it has been proved by several authors that the representation of the data can play a crucial role in the learning capabilities of an NN (see e.g. Ref. 28).

One of the main problems, associated with the learning phase, is the risk of overtraining (see for example Ref. 29). If the examples have some (non-evident) features which are not shared by the represented class, the system, after a certain number of iterations of the training procedure, will start to include those features into the model, thus deviating from the desired generalization process. Roughly speaking, the problem of overfitting is associated with an excessive influence of the examples onto the model, that is, onto the parameters of the system. Sjoberg and Ljung[29] said, "In the case of a neural network trained with a gradient method, overfitting may happen. However, it does not come instantaneously. Instead the overfitting comes gradually during the training, i.e. if the *error* function is computed on independent data after every training cycle, it will first decrease and the network learns to approximate the unknown function, and then it will start to grow again ...".

As a consequence of their analysis, we can assert that when the model is sufficiently flexible to include the correct parametrization of the desired function, each estimated weight contributes to the error. But when the NN is over-parametrized, which means that it has too many weights to be tuned, it can give rise to an overfitting, due to the fact that some parameters do not improve the error function but make the model worse when applied to a new data set. With $\mathcal{D}_1$ representing the trained data, $\mathcal{D}_2$ representing a set of *untrained data*, f_1 identifying the model with superfluous parameters and f_2 the model without them, this can be interpreted as follows:

$$E[(y - f_1(x; \mathcal{D}_1))^2 | x, \mathcal{D}_1] \approx E[(y - f_2(x; \mathcal{D}_1))^2 | x, \mathcal{D}_1]\,,$$

but, for the corresponding variances,

$$E_{\mathcal{D}_2}[(f_1(x;\mathcal{D}_2) - E_{\mathcal{D}_2}[f_1(x;\mathcal{D}_2)])^2] \gg E_{\mathcal{D}_2}[(f_2(x;\mathcal{D}_2) - E_{\mathcal{D}_2}[f_2(x;\mathcal{D}_2)])^2]\,.$$

A good remedy against overfitting involves a suitable reduction of the free weights of the system. This can be done, for example, by techniques which bias the system through appropriate *a priori* properties or through statistical methodologies like smoothing. The last is most appropriate when designers have no ideas concerning the exceeding parameters.

Interestingly, Sjoberg and Ljung have shown that interrupting the minimization procedure before convergence has the same effect of a regularization process on the data, inducing a reduced variance to the model. Using this as a real-world industrial application, they had studied the modeling of a robot arm, on which they successfully experimented their results.

Based on this analysis, we study in the next section a specific industrial problem of optical inspection on the surface of metallic parts.

4. An Industrial Application

In this section we discuss a general methodology for neural networks design, which aims to drastically reduce the degrees of freedom of the system by exploiting the *a priori* knowledge of the problem, and also to overcome the limits caused by a small number of available examples. Next, we discuss a specific application proposed by the authors which addresses a "crack recognition" problem. The resulting network accepts images of parts under inspection at its input and issues at the output a flag which states whether or not the part is defective.

The results obtained so far show that such a classifier provides a potentially relevant approach for the quality control of metallic objects since it offers at the same time high performance and a moderate effort in software development.

4.1. *The input pattern*

The problem we consider consists in the image inspection of the mechanical machined parts. Our aim is to identify the defective parts, and the suggested procedure works towards the goal of isolating possible cracks by highlighting them with respect to the scene. The problem can be formulated as a binary classification problem, that has to be integrated into the production process, ensuring high classification rates, good flexibility and fast performance.

The input data consist of 256 gray-level images having a size of 740×560 pixels. These images are obtained using a UV-sensitive camera positioned along a production line. We will further discuss about this consolidated methodology.

Visual inspection of industrial machined parts is a widespread technique, having an important support in non-destructive magnetic methods,[30–32] which allows us to identify possible defects present on the surface of metallic objects, without altering their properties. In our application, we have to do with an FMPI (fluorescent magnetic particle inspection) treatment, which exploits the leakage of magnetic flux in the neighborhood of surface discontinuities.[33] The application of an external, sufficiently intense, magnetic flux yields a redistribution of magnetic particles, previously deposed on the metallic surface. The main effect is visible near surface discontinuities, where one can observe the density maxima of the magnetic powder.

By using fluorescent magnetic particles, such redistribution is made evident by an acquisition of the image under ultraviolet lighting. This makes the irregularities of the surface clearer, and helps to identify structural defects. Some examples are illustrated in Figs. 3 and 4.

Remark 4.1: *Since the magnetic field on the powder gives rise to complex and noisy image patterns, an algorithm based on thresholding cannot identify the crack.*

A possible "mechanical" pre-processing is suggested by Cheu[34] in his project on an automated connecting-rod crack detection system, which involves a controlled rinsing of the part to weaken the effects of the edges which are clearly false defects. Unfortunately, in the presence of complex geometries, this mechanism is not reliable enough and the required rinsing is difficult to accommodate in several manufacturing steps. For this reason, a robust algorithm overcoming the previous limitations would provide a more effective inspection technique. The system we propose does not require any mechanical preprocessing of the image and is thus easier to use in a standard environment.

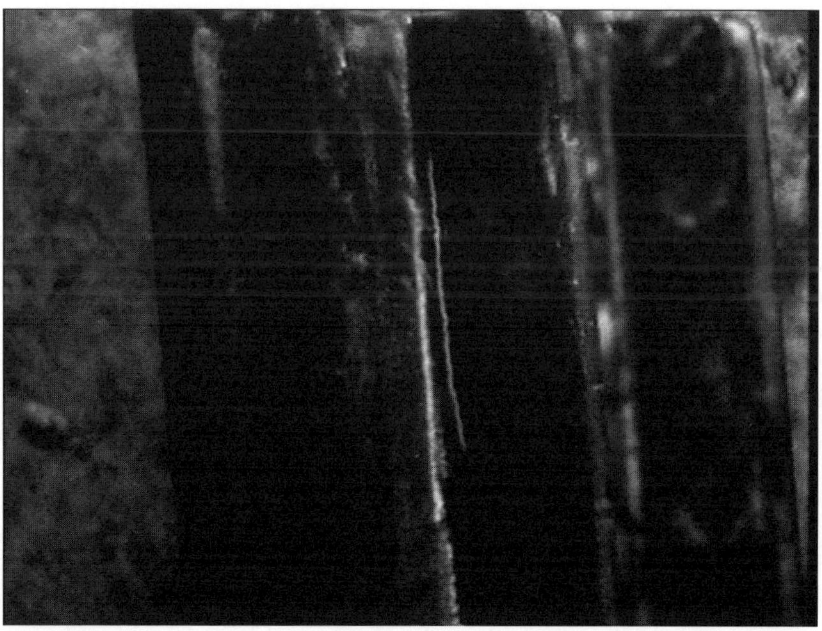

Fig. 3. An example of image with an elongated crack.

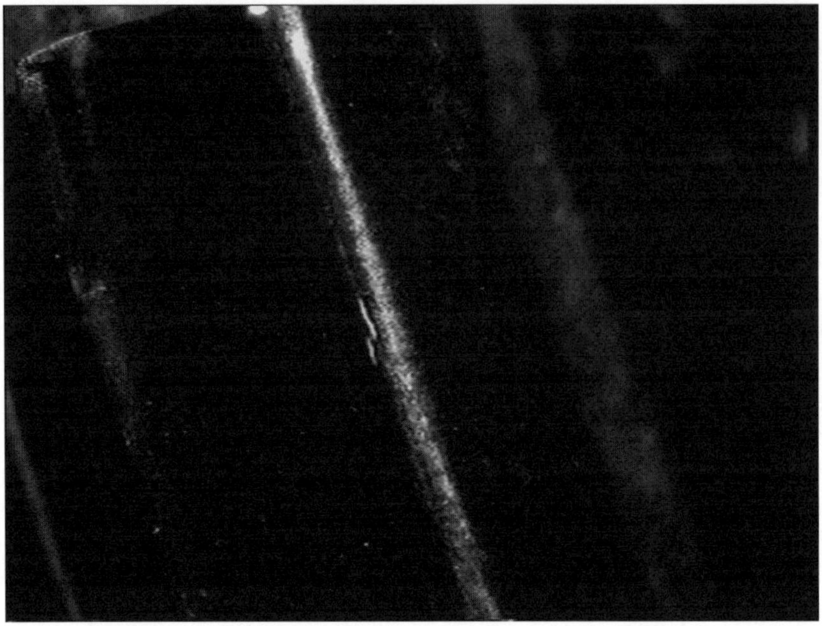

Fig. 4. An example of image with a localized flaw.

4.2. *The architectural choice*

As previously mentioned, we opted for a strongly-constrained architecture, since a substantial amount of *a priori* information that must be included in the classification model is available. Hence, the system development began with interviewing the experts who provided a substantial number of rules in terms of what is *not* relevant to the task at hand. This suggested the recast of this information in terms of a suitable set of invariances.

In summary, from a structural point of view, we aim to design a strongly predefined network. However, we also require the system to be able to adapt its parameters to the moderate variations in the application features.

The design of a strongly biased network gives rise to several important features. First of all, with reference to the small dimensions of the training database, we move towards a solution characterized by a small *variance* (see Sec. 2). Next, we get a fast convergence of the learning phase and a better control of the optimization process, reducing the computational effort (as we shall see in Sec. 4.5.2).

4.3. *Biasing the network*

A first hint provided by the experts is that they are able to detect a defect just by looking at a portion of the image. A qualitative bound on the required image size is about 40×40 pixels. This fact implies that it is possible to examine relatively small portions (possibly overlapping) of the image *in parallel.* Another powerful constraint is that since the defect is a crack, its shape is roughly one-dimensional. Even though its length can be substantial with strong variations, its thickness does not exceed a maximum of 9 pixels in our camera setup. Furthermore, a direct inspection of the cracks shows that their local structure is approximately linear, even though the noise can introduce relevant distortions to their global aspect. This suggests the use (at the lowest level) of the network of local feature detectors associated with square blocks having a side somewhat larger than the thickness of a crack. These blocks, possibly overlapping, cover the portion under consideration of the image. The output of the feature detectors can then be fed to a network which combines all the available outputs to provide the final classification. In this framework, spatial invariances directly restrict the mutual relationships among weights values of the feature detectors. More specifically, the classifier, and thus the feature detectors, should be invariant to translations, rotations and chirality inversion of the image, while the scale invariance is not applicable to this problem. The first two invariances, derived from a natural principle of isotropy, have been investigated by others (e.g. Ref. 35), obtaining networks that are too complex for our application. For the more manageable networks only rotation invariance[36] and approximate translation invariance[37] have been presented, while chirality invariance has not received any attention as yet.

The main methods in obtaining a network with a behavior invariant to a set of spatial transformations[35] are based either on a suitable training procedure, or on a choice of an input-data representation which embeds into itself the invariances or, finally, based on the determination of a proper network topology. The first choice implies a learning of invariances by examples, which is possible only when a large number of training samples is available. Since the information about the invariances is not explicitly exploited, the size of the network remains very large, leading to high-variance solutions. The second choice involves the definition of representation systems of the input data having the invariance built in. For example, a polar parametrization of linear patterns automatically provides rotation invariance to the following stages of a network. This strategy is very powerful, but generally it requires a high computational effort to project the input data onto the invariant features space. The third technique is based on the possibility of designing a network topology embedding the required invariance. This technique is used in this work since it reduces substantially the number of free parameters of a network and it lets the learning procedure determine the best parameters required by the process.

4.4. *Architecture of the network*

The classifier examines subimages having the size of $L \times L \sim 40 \times 40$ pixels and is consisting of three layers.

We now describe the architecture of each layer of the network which is represented in Fig. 5, starting from the input.

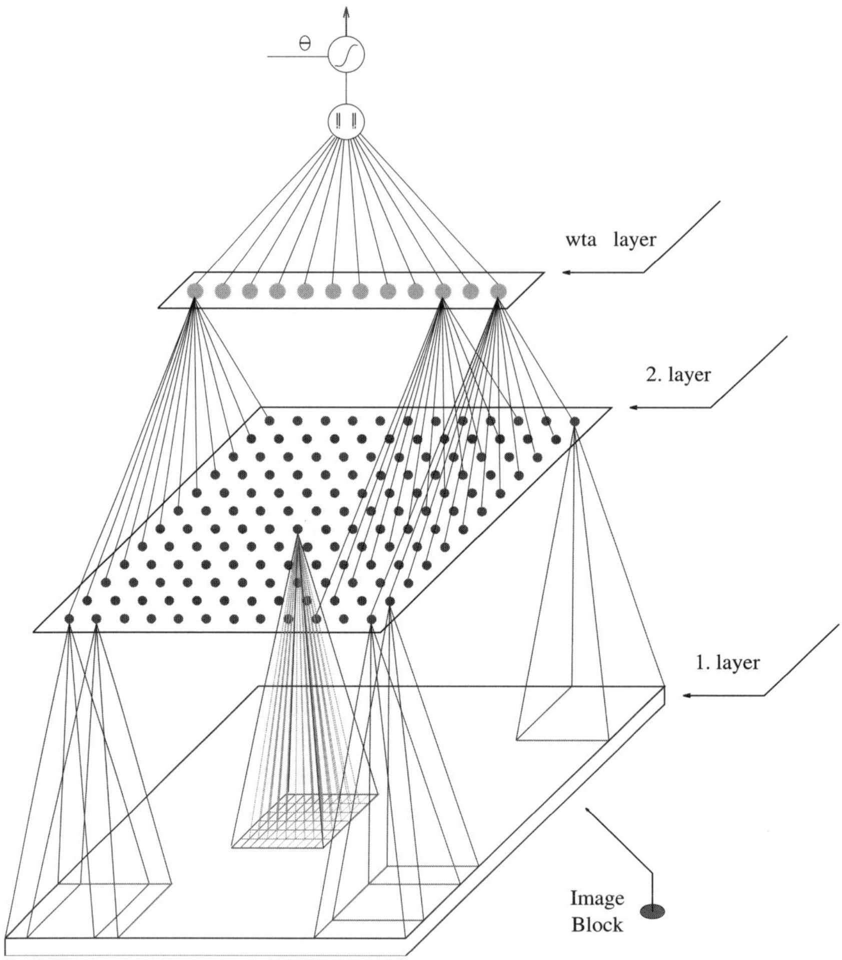

Fig. 5. Architecture of the global NN.[3]

4.5. *An adaptive filtering of the image*

The units in the first layer act as feature detectors tuned to the shape of the defects. In other words, they provide a suitable filtering of the image. For the sake of simplicity, we describe an architecture with a single family of feature detectors in this paper.

4.5.1. *The feature detectors*

The general form of the feature detector f that we consider is:

$$f(j,i) = \sum_{(u,v)\in S_{ji}} \mathcal{F}_{ji}(u,v) \cdot \mathcal{I}(j-u, i-v)\,, \tag{3}$$

where $\mathcal{I}$ is the discrete input image, (j,i) addresses the generic pixel and $\mathcal{F}_{ji}$ is a discrete convolution kernel related to pixel (j,i) and having a suitably chosen compact support S_{ji}. If several detectors are used at the same time, each of them must satisfy the constraints as discussed in the previous section, providing a linear increase in the number of independent weights. We shall now describe how the different invariances restrict the weights values of the feature detectors.

Translational invariance. Each unit covers a square patch of the image having size $N \times M$ that is of the same order as the thickness of the crack (about 5×9 pixels in our implementation) and shares the value of its weights with all the other units. Thus it achieves positional invariance,[37] and provides a strong reduction of free parameters. These, possibly overlapped, patches completely cover the image. Hence Eq. (3) can be simplified as:

$$f(j,i) = \sum_{n=-N/2}^{N/2} \sum_{m=-M/2}^{M/2} \mathcal{F}(n,m) \cdot \mathcal{I}(j-n, i-m)\,, \tag{4}$$

where $\mathcal{F}$ is independent of (j,i).

N and M are chosen as odd numbers for symmetry requirements and the quotients $N/2$ and $M/2$ are integer divisions, i.e. $N/2 = (N-1)/2$, and similarly for M.

Structural invariance. The number of weights can be further reduced by checking that a locally linear structure is obtained. Hence, the weights must share the same values along the direction orthogonal to the crack. This reduces the number of independent weights by a factor of N. However, the dimensions of the processed patch remains at $N \times M$ in order to filter out the noise of the image. This is carried out by the low-pass behavior of the detector along the direction parallel to the crack. Thus, we obtain from Eq. (4):

$$f(j,i) = \mathcal{C} \sum_{n=-N/2}^{N/2} \sum_{m=-M/2}^{M/2} \mathcal{F}^{\ominus}(m) \cdot \mathcal{I}(j-n, i-m)\,,$$

where $\mathcal{C}$ is a suitable constant.

Chirality invariance. Finally, chirality invariance requires the weights associated with the pixels on the left side of the patch to have the same value as those on the right side of the patch. This means that $\mathcal{F}^{\ominus}$ has a central symmetry with respect to the midpoint of its support. Thus, the total number of independent weights required by the feature detectors decreases now to $(M/2)+1$ (5 in our implementation). More in detail, we obtain:

$$f(j,i) = \mathcal{C} \sum_{n=-N/2}^{N/2} \sum_{m=0}^{M/2} \mathcal{F}^{\ominus}(m) \cdot (\mathcal{I}(j-n, i-m) + \mathcal{I}(j-n, i+m))\,. \tag{5}$$

Rotational invariance. Approximate rotational invariance is obtained by scanning multiple directions along the same image. This process can be formally described considering a number, D, of images I_{α_d}, where $d = 1, \ldots, D$ is derived from the same scene through the rotations of angles α_d. A possible choice for D is $D = 2^k$, with k as a positive integer.

In our experience, four images rotated at 45°, result in $k = 2$ which is a reasonably approximate rotational invariance. We can associate every image I_{α_d} with a network whose structure is independent from the processed direction and shares its weights with the networks associated with the other directions. Note that a better angular resolution increases the computational load, while the number of independent weights remains the same. Hence, the number of operations increases linearly with the angular resolution, while the robustness of the classifier, as determined by the number of free weights, is not affected.

4.5.2. *Improving the computational efficiency*

Now, we will have a short discussion concerning an implemented procedure which speeds up global processing. Since the assumed invariances are quite general, such analysis could easily be extended to other applications. By examining Eq. (5), one readily gets:

$$f(j+1,i) = f(j,i) + \gamma_d - \gamma_u \,,$$

where

$$\gamma_d = \mathcal{C} \sum_{m=0}^{M/2} \mathcal{F}^{\ominus}(m) \cdot \left(\mathcal{I}\left(j+1+\frac{N}{2}, i-m\right) + \mathcal{I}\left(j+1+\frac{N}{2}, i+m\right)\right),$$

$$\gamma_u = \mathcal{C} \sum_{m=0}^{M/2} \mathcal{F}^{\ominus}(m) \cdot \left(\mathcal{I}\left(j-\frac{N}{2}, i-m\right) + \mathcal{I}\left(j-\frac{N}{2}, i+m\right)\right).$$

Consequently, the global computation may be implemented by *incremental* steps.

By defining a matrix γ, whose elements are defined by:

$$\gamma_{\ell,k} := \sum_{m=0}^{M/2} \mathcal{F}^{\ominus}(m) \cdot (\mathcal{I}(k, \ell-m) + \mathcal{I}(k, \ell+m)) \,,$$

and by setting

$$\mathbf{G}(i,j) := \gamma_{j+1+N/2,i} - \gamma_{j-N/2,i} \,, \tag{6}$$

one gets

$$f(j+1,i) = f(j,i) + \mathbf{G}(i,j) \,. \tag{7}$$

Thus we can propose the following algorithm to compute the output of the convolution filter at the first layer level on the whole image (which yields most of the computational effort):

(i) Compute the matrix γ, by performing the convolution $\gamma = \mathcal{F}^{\ominus} * I$;
(ii) Compute the first row of the filtered image f;
(iii) Compute the following rows of f using Eqs. (6) and (7).

This way, when passing from one patch to the following, part of the previous computation of the NN may be saved and shared with the next pattern, providing a remarkable speed up.

4.5.3. *Hidden layers of the network*

The internal layers of the network gather the evidence provided by the feature detectors and then perform the final classification of the image. Since we are seeking an almost linear structure, the layer collecting the output of the feature detectors is organized as a set of D groups of Winner-Take-All (WTA) networks, with one for each angle α_d. Considering a direction parallel to the columns of matrix I_{α_d}, the associated group of WTAs selects for each row of the matrix $f_{\alpha_d} = \mathcal{F} * I_{\alpha_d}$ the maximum of the values and stores them in a vector $V_{\max}^{\alpha_d}$:

$$V_{\alpha_d}^{\max}(l) = \max_{i=1,\ldots,L} \{f_{\alpha_d}(l,i)\} \quad l = 1,\ldots,L\,.$$

Note that, given a certain direction α_d and the related image I_{α_d}, the associated feature detector is characterized by a structural symmetry axis parallel to the direction under investigation. Thus, the following group of WTAs processes the filtered image f_{α_d} along a direction orthogonal to the one scanned by the feature detector. Each WTA is then organized as a binary tree of modules computing the "soft" maximum of two input values.

Differentiable WTA

In order to use the error back propagation technique[20] it is necessary to consider a differentiable approximation of the *maximum function.*

We use an approximation of the Heaviside distribution through a suitable pointwise convergent sequence of sigmoidal functions. If $H(x,z)$ is the shifted Heaviside function with discontinuity at z, this is done as follows:

$$H(x,z) \approx \sigma_n(x,z) = \frac{1}{1 + \exp\left(-n(x-z)\right)}.$$

We easily get:

$$\lim_{n\to\infty} \int_{-\infty}^{\infty} |\sigma_n(x,z) - H(x,z)|dx = 0\,.$$

After a suitable choice of n, we set the following approximation:

$$\max(x,z) \approx \int_{-\infty}^{\infty} \sigma_n(x,z)\,dx + z = \frac{1}{n}\log\left(\exp\left(n(x-z)\right) + 1\right) + z\,.$$

Then, with some algebraic manipulation, we get the following monotonic differentiable soft maximum function:

$$\max(x, z) \approx -\frac{1}{n} \log(\sigma_n(z, x)) + z \tag{8}$$

whose steepness can be tuned by changing the value of n. Note that this step does not introduce any free parameter.

4.5.4. *The output layer*

The final classification is then performed by a set of sigmoidal units in the third layer, which is associated to each vector $V_{\max}$ produced by the previous blocks. Each unit at this level estimates the norm $N(\alpha_d)$ of the related vector $V_{\max}^{\alpha_d}$ and then produces a flag $C_d(\alpha_d)$ at its output:

$$C_d(\alpha_d) = \sigma(N(\alpha_d) - \Theta)\,, \qquad d = 1, \ldots, D\,, \tag{9}$$

where Θ is the neuron threshold learnt during the training phase and σ is a sigmoidal function. If at least a flag $C_d(\alpha_d)$ is issued, then a crack is detected in the image. Since all units have a differentiable transfer function, standard error back propagation can be used to determine the weights and thresholds of the network.

4.6. *Results*

The classifier has been implemented using Aspirin,[38] a NN development tool. To train the classifier over a set of 200 independent subimages each having a size of 40×40, all suitably extracted from three images, and randomly picked out from a group of 20 images, we made use of an EBP algorithm, implemented both in the traditional way and having some variants (see for example Ref. 39). Ten patterns showed cracks, while 190 were associated with a normal surface. The large number of examples associated with good parts reflects the relative scarcity of production defects. However, since in this application the cost of *not* detecting a crack is potentially high, we made the examples of the crack in the training sequence as frequent as the examples of the normal parts. This procedure requires the use of a suitably weighted Bayes decision rule.[22] We have then tested the performance of the classifier on a set of 3000 patterns extracted from the remaining images. Figure 6 shows the profile of the weights of the first layer, after the learning phase.

The network classifies images (such as the ones shown in Figs. 3 and 4) and provides continuous values for the probability of having a defect in the region under consideration. A typical output is shown in Figs. 7 and 8, where the probability of having a defect is encoded using gray levels.

Note that only the real crack has been highlighted, even though other structures with higher brightness and elongated shape are visible in the scene. Figure 6 shows that the weights learnt by the feature detectors capture the shape of a crack. The obtained filter exhibits a differential behavior, showing that high frequency

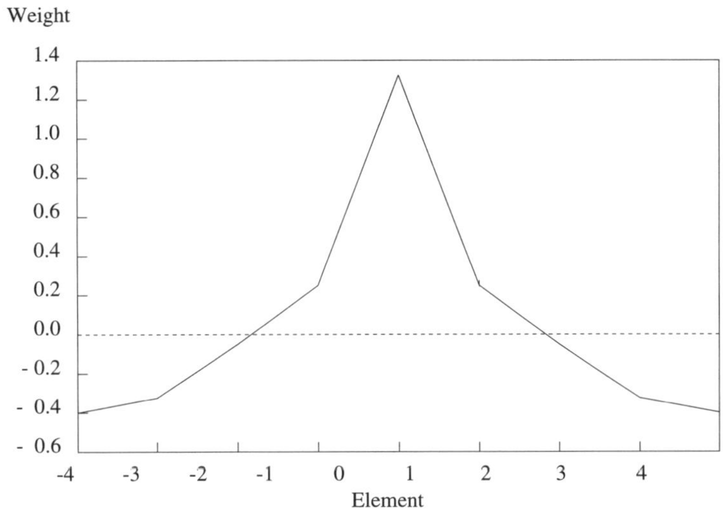

Fig. 6. Profile of the main feature detector.

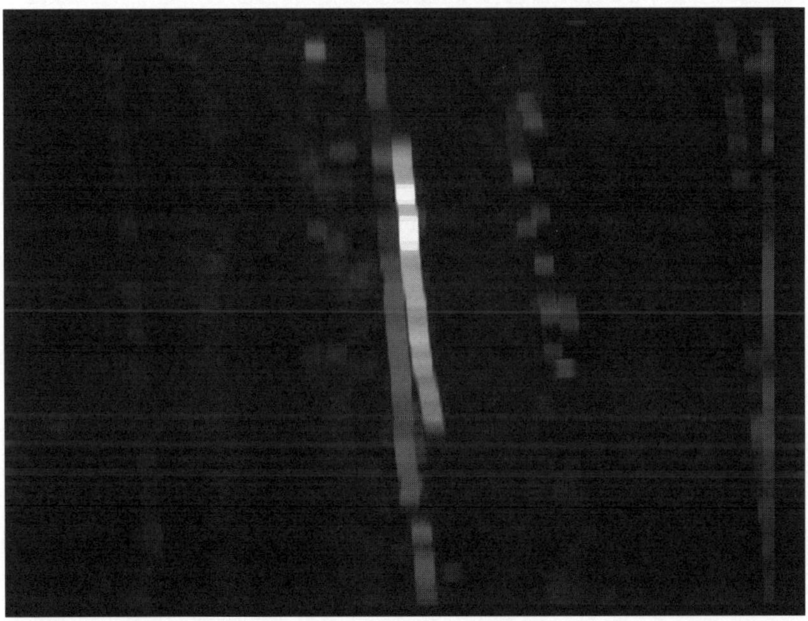

Fig. 7. Output pattern for the image in Fig. 3.

components of the analyzed signal are particularly meaningful. The error has been evaluated using a cross-validation technique. The CPU time required by the learning phase is a few minutes on a SPARC 20 Sun Workstation. Such a short CPU time has been obtained since the number of weights to be learnt is extremely small.

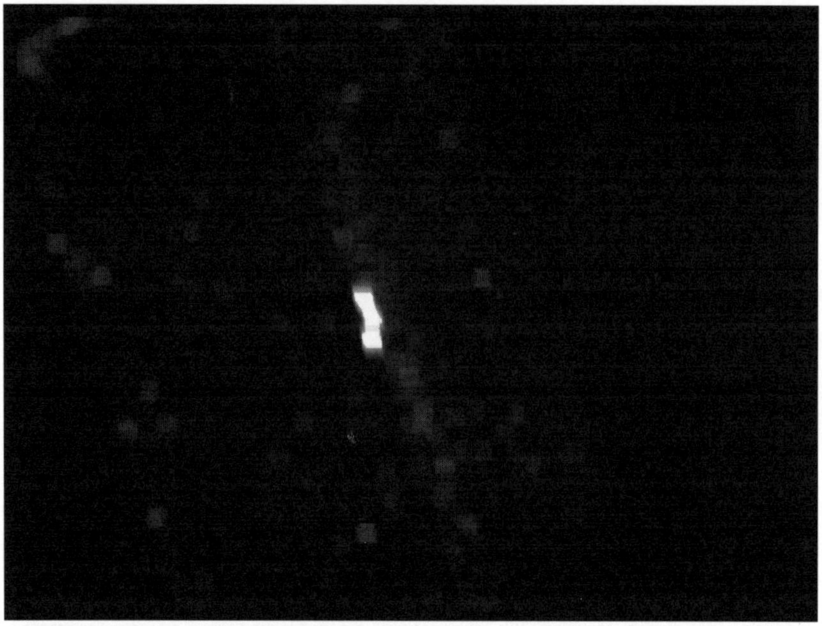

Fig. 8. Output pattern for the image in Fig. 4.

Moreover, the error on the test set has been experimentally shown to be quite insensitive when the error on the training set is made negligible. This implies that the network does not exhibit the overfitting problem typical of redundant classifiers.[9] Furthermore, a robust property of the network is illustrated through the investigation of the reduction effect of the gray levels. In Ref. 3 it is shown that the error does not appreciably increase when reducing the number of bits used to code the gray levels of the image down to three bits.

Finally, we have studied the effect of a lower resolution on the frequency domain. This experiment was carried out by defocusing the lens of the camera. This is equivalent to a low-pass filtering of the image. The loss of the high frequency components introduces a serious degradation of the classifier performance, confirming that it behaves like a frequency-tuned system. However, note that a blurred image is unusual in industrial applications where the camera stands still and its setup can be calibrated with high precision.

4.7. *A hardware implementation*

In Ref. 40 a 0.7 μm CMOS ASIC chip that implements the neural network previously described for real-time image processing has been proposed. The chip, which is a module of the global system for the automatic surface inspection of mechanical parts, implements the feedforward phase of the neural-network model. The architecture is based on a deep pipeline and its performance exceeds real-time specifications.

According to our experiments and the analysis performed in Ref. 40, the computation precision needed by the neural architecture shows that the multiplications involved in the computation of the convolution (at the first layer level) can be efficiently performed with four to five bits. Consequently, the multipliers have been designed using look-up tables, as suggested by Ref. 41. The implementation was carried out using RAMs.

The chip consists of 13 000 standard cells and 7 RAM macroblocks; the chip area is 10×0.9 mm. The measured speed is about 30 MHz corresponding to 1.35 GOPS. The power consumption is about 2 W at a supply voltage of 5 V.

4.8. *Additional remarks*

In short, we have realized a contextual adaptive network, which achieves a high reliability in detecting a type of pattern (the cracks) within extremely-variable scenes. In our application, most of the information has been stated in terms of spatial and structural invariances and has been used to bias the network towards an acceptable solution. There are cases, however, where a local analysis of the scene is not sufficient to provide a correct classification (this is the case of a milled cam shaft, which we met in one of our tests). In this case, some experts asserted that a correct interpretation required a contextual analysis of the whole image, and thus could not be obtained by the pure analysis of small patches. With reference to this case, the approach we propose must be considered effective only if it does not require the general context of the scene.

In most cases, however, we have found that a local analysis of the image is sufficient for a correct classification.

5. Second Application: On-line Inspection of Surface-mount Devices

We describe here a methodology for the on-line inspection of surface-mount components, as proposed by Dar *et al.*,[42] which is oriented to the quality analysis of printed circuit boards.

In the last few years the need for quality assurance of manufactured printed circuit boards has increased. Dar *et al.* said, "... many studies have been performed that show, on a given day, that an operator may inspect the same printed circuit board and declare different defects. This lack of reliability and repeatability demands a more accurate approach using automated systems". The first stage of the method which they propose is based on visual inspection and infrared sensors. This has the aim of separating the solder joint defects into surface-level defects and solder-mass related defects before the final detection. The complementarity of infrared and vision sensors improves the quality of the inspection strategy, but the presence of multiple sensors requires a complex system that is able to exploit all the information available to provide the final classification. Since the system has to be adaptive, it needs the provision of suitable feedback mechanisms. Finally, to make the system effective, the cost has to be limited and real-time operating-conditions must be guaranteed.

Since the quality control requirements for printed circuit boards can vary from one industrial application to the other, one has to find a compromise between robustness, accuracy and flexibility. There is in fact no universal definition of a good solder joint, due to the large variability of applications and products which leads to different quality requirements.

The proposed inspection methodology comprises of two stages: the first stage, called the "GROSS inspection station", scans a printed circuit board with the goal of extracting global features and determining the candidate defective areas of the board. The second stage, called the "FINE inspection station", provides a more accurate (but local) analysis of the previously extracted areas. After the local features extraction, the system provides the final classification. Since the second stage of the inspection is developed off-line, the authors obtained a time-efficient global inspection system. Furthermore since this stage is implemented by standard pattern-analysis techniques, we only describe here the first stage, which has to be run on-line.

5.1. *The system*

The system under consideration includes both standard and infrared sensors. The vision sensors are gray scale CCD which produce 256 gray-level images with 640×480 size. The visual inspection consists of image processing and subsequent pattern analysis for the detection of visible defective joints.

The infrared inspection makes use of a laser which heats the joints and allows the infrared radiance curve to be analyzed as the joints heat up and then cool down. A two-color infrared sensor is used to record the radiance curve of the solder joints (for the details, see Ref. 43).

First, the system is employed on-line to detect missing components, angular misalignments, linear misalignments and solder-mass defects. If a candidate defective board is found, it is shipped to the second module of the system (FINE), where another vision inspection is performed, which provides a higher-resolution 256 gray-level image, by simply zooming in on the possibly-defective area. As for the infrared analysis, after the phase where the defects have been highlighted, a statistical sampling scheme is used for the infrared screening phase. This was performed with the goal of detecting defect categories such as excess solder, insufficient solder, cold joint and so on. In such a phase, only the peak of the radiance curve is considered in order to label the part as "possibly defective". Then, passing on to the second module, the complete radiance curve of a candidate defective solder joint is analyzed in detail, which means that both the heating and the cooling cycles are finely examined. After this, every possible defect is definitely classified.

This way, the time required to inspect the joints is reduced; such reduction can be tuned to the specific tasks, according to its constraints, by adjusting the sampling rate.

Typically the solder joint is brighter. This allows us to extract it from the backround by simple thresholding. Furthermore, to obtain a higher absorption of the incident light, a red-pass filter is used for the illumination. Moreover, a circular optical-fiber ring light is used to enable a uniform, shadow-free illumination. Both the modules are controlled by a PC.

5.2. *The GROSS module*

The first step in visual inspection consists of detecting the presence of all components on the board. After this, the system checks both the angular and the linear alignment of the components. To do this, two images are captured; the first one under oblique illumination for testing the presence of components; the second under a flat illumination for testing possible misalignments.

An important feature of the module is that it is designed to provide a non-destructive inspection by ensuring that no defective solder joints are declared to be good.

5.2.1. *Component presence: neural based classification*

To detect missing components, the system illuminates the board by projecting a goose-neck halogen light source at 30° from the board plane. This way, if a component is present, a shadow is cast.

By using computer aided design informations, which are available in the production process, it is possible to associate suitable windows to the shadow-side of all components. This is commonly done during the training phase of the adaptive system.

Every window identifies a sub-image, whose gray-scale histogram is computed and stored. Let H be the histogram of an $m \times n$ window. To ensure a normalized representation, independent of the window size, it is sufficient to set the normalized histogram as $H^\star := H/(mn)$. Examples of such histograms are illustrated in Ref. 42. By extracting some statistical features from the histogram of a component, e.g. the expected value and the standard deviation, it is shown that linearly separable classes are obtained. Consequently, a simple neural classifier, with a small number of free parameters, has been implemented on a training set consisting of 34 examples, from which six were randomly extracted to provide a validation. This allows us to control the variance of the classifier with respect to the training set, as discussed in Sec. 3.

5.2.2. *Linear misalignment: fuzzy-neural based classification*

As anticipated, a red-pass filter is added to the light source providing an enhanced contrast flat illumination (this is illustrated by Ref. 42 in Fig. 6). Based on computer aided design informations, programmed rectangular windows have been designed to obtain sub-images of the solder joints of the components. From every sub-image a vector is extracted, whose peaks represent solder joints and solder pads, while intervalleys represent the substrate. This vector is obtained by a thresholding based segmentation of the sub-image. In particular, starting from a sub-image I_ν of size $m \times n$, $(m \geq n)$, the jth element of the vector, say v_ν is determined by means of a WTA layer, as

$$v_\nu(j) = \max_{k=1}^{m'} A_\nu(k, j)\,,$$

where $m' = m - q$ and $A_\nu(k, j)$ is the average of the kth set of q pixels in the jth column of the sub-image I_ν, that is

$$A(j, k) = \frac{1}{q} \sum_{i=1}^{q} I(k + i, j)\,.$$

The subsequent step is obtained by applying a morphological filtering process to the computed vector to remove the noise, based on the morphological dilation and erosion operators (see for example Ref. 44).

Finally, a connectivity analysis is performed to detect the length of the dark and of the bright regions extracted. The length of the bright region determines the width of a solder joint, while that of a dark region determines the distance between two adjacent solder joints. To determine misaligned solder joints, a fuzzy classification technique has been employed. Using the width of the solder joint and the distance between two adjacent solder joints as features, a training set was created consisting of a set of patterns partitioned into a number of classes representing different surface mount components to be inspected. In this framework, a fuzzy classifier was implemented based on fuzzy relations constructed using the *a priori* information available for a set of good and a set of defective elements. See Ref. 42, Sec. 4 for a detailed description of this classifier.

5.2.3. *Angular misalignment: adaptive thresholding*

The sub-image used in the previous analysis is processed again to determine the outside corners of the first and the last solder pads. Similarly, the outside corners of the component leads are determined. Then the slopes s_p and s_l of the two lines connecting the corner pad and the component lead are computed. The difference between the two slopes provides the angular error $\Delta_\theta = |\tan^{-1}(s_p) - \tan^{-1}(s_l)|$, which is compared to a fixed threshold to be adaptively determined. This provides the possible angular misalignment of the pads with respect to the component leads.

5.3. *The infrared based inspection*

This second functionality of the "GROSS inspection system" determines the presence of solder mass defects. The analysis is based on the dependence of the infrared signal on the radiance of the component, which is affected by the amount of mass. For the defects related to a lack of mass, the thermal signature turns out to be elevated due to the increase of the infrared radiation (the heating time is assumed to be fixed). For the defects related to an excess of solder, the thermal signature is decreased.

The following set of features was extracted from the thermal signatures:

(i) the maximum of the signature;
(ii) the slope of the heating curve (from the initial point to the maximum);
(iii) the maximum deviation from the slope of the curve;
(iv) the temperature at the boundary;
(v) the area subtended by the curve; and
(vi) the frequency spectrum of the curve.

The processing flow starts off-line, where the location of the component solder joints are programmed and stored into the system. Then, the thermal signatures are recorded on line at the preprogrammed locations. Finally, the features of the signals are extracted. In order to improve the quality of the signal, which can be very noisy, a low-pass filtering is performed.

The average time required for the inspection of a solder joint is about 500 ms. The "GROSS module" makes use of a subset of the extracted features. On this basis, it labels the joints and possibly supplies them to the second module for a further inspection. This time is based on the complete feature set.

Based on the proposed features, an efficient defect classification is achievable, for example, by using a two layer perceptron.

For what concerns the detailed description of the second module (FINE) and its technical details, we address the reader to the papers by Dar *et al.*

5.4. *Results and remarks*

The system was tested on a set of five Ethernet cards. The neural classifier developed to check the component presence provided 100% correct classifications. Moreover, the whole classifier showed robustness with respect to the illumination. The subsystems devoted to the linear and angular misalignments found all the existing defects, while the false alarm rate amounted to 3–4% of the "real" defects.

The combination of visual and infrared inspection provides a high computational efficiency to the whole detection system, which is also improved by the hierarchic processing inherent in the pipeline architecture comprising of the "GROSS" and the "FINE" modules.

6. Conclusions

In this paper we have presented a short survey of neural-network techniques for the optical surface inspection of machined parts. Following that we discussed the theoretical framework, describing some general problems related with these applications and then stressing the importance of exploiting knowledge-based properties in terms of structural invariances.

Finally, with reference to a couple of specific applications, we have investigated techniques for embedding the aforementioned domain-specific informations into constrained adaptive networks. As shown by the examples, this information is used to drastically reduce the number of free parameters which must be determined during the learning phase, thus allowing artificial neural networks to be applied to problems characterized by a relatively small number of available examples.

Acknowledgments

This work has been partially supported by ST-Microelectronics under the National Programme on Bioelectronic Technologies.

References

1. C. Neubauer, Fast detection and classification of defects on treated metal surfaces using a backpropagation neural network, *IJCNN-91, Singapore* **2** (1991) 1148–1153.
2. C. Wang, D. J. Cannon, S. R. T. Kumara and G. Lu, A skeleton and neural-network based approach for identifying cosmetic surface flaws, *IEEE Trans. Neural Networks* **6**, 5 (1995) 1201–1210.
3. N. Guglielmi, R. Guerrieri and G. Baccarani, Highly-constrained neural networks for industrial quality control, *IEEE Trans. Neural Networks* **7** (1996) 206–213.
4. R. T. Chin, SURVEY: automated visual inspection: 1981 to 1987, *Computer Vision, Graphics, and Image Processing* **41** (1988) 346–381.
5. P. H. Winston, Learning structure descriptions from examples, ed. P. H. Winston, *The Psychology of Computer Vision* (McGraw Hill, New York, 1975) 157–210.
6. F. Tomita, Interactive and automatic image recognition system, *Machine Vision and Application* **1** (1988) 59–69.
7. N. Guglielmi and R. Guerrieri, An experimental comparison of software methodologies for image based quality control, *IEEE Industrial Electron. Soc., Proc. IECON-94* **3** (1994) 1942–1945.
8. R. P. Lippmann, Pattern classification using neural networks, *IEEE Communication Magazine* (1989) 47–64.
9. S. Geman, E. Bienenstock and R. Doursat, Neural networks and the bias/variance dilemma, *Neural Computation* **4** (1992) 1–58.
10. T. S. Newman and A. K. Jain, CAD-based inspection of 3D objects using range images, *IEEE Industrial Electron. Soc., Proc. IEEE Workshop CAD-Based Vision*, Champion, Pensylvania (1994) 236–243.
11. C. Bradley and S. Kurada, Industrial inspection employing a three dimensional vision system and a neural network classifier, *IEEE Industrial Electron. Soc., Proc. PRCCCSP-95*, Victoria, Canada (1995) 505–508.
12. M. A. Sid-Ahmed, J. J. Soltl and N. Rajendran, Specific applications of image processing to surface flaw detection, *Comput. Industry* **7** (1986) 131–143.
13. H. Blum, A transform for extracting new descriptors of shape, ed. W. Wathen-Dunn, *Models for the Perception of Speech and Visual Forms* (MIT Press, Cambridge, MA, 1967).
14. Z. Zhou and A. Venetsanopoulos, Morphological skeleton representation and shape recognition, *IEEE Industrial Electron. Soc., Proc. ICASSP-88* (1988) 948–951.

15. A. König, H. Genther and M. Glesner, Neural and associative modules in a hybrid dynamic system for visual industrial quality control, *IEEE Industrial Electron. Soc., Proc. ICNN-93* **3**, San Francisco, CA (1993) 1510–1515.
16. W. Poechmueller, M. Glesner, L. Listl and P. Mengel, Automatic classification of solder joint images, *IEEE Industrial Electron. Soc., Proc. IJCNN-91* **2** (1991).
17. H. H. Szu, Automatic fault recognition by image correlation neural network techniques, *IEEE Trans. Industrial Electron.* **40**, 2 (1993) 197–208.
18. H. H. Szu and J. Blodgett, Self reference spatio-temporal image-restoration technique, *J. Opt. Soc. Am.* **72** (1982) 1666–1669.
19. L. Villalobos and S. Gruber, A system for neural network-based inspection of machined surfaces, *J. Neural Network Computing* **2**, 2 (1990) 18–30.
20. D. Rumelhart, J. L. McClelland and the PDP-group, *Parallel Distributed Processing* **1** (MIT Press, 1986).
21. S. A. Solla, Learning and generalization in layered neural networks: the contiguity problem, eds. L. Personnaz and G. Dreyfus, *Neural Networks: From Models to Applications* (IDSET, Paris, 1989) 168–177.
22. K. Fukunaga, *Statistical Pattern Recognition* (Academic Press, San Diego, CA, 1990).
23. R. O. Duda and P. E. Hart, *Pattern Classification and Scene Analysis* (Wiley, New York, 1973).
24. T.-H. Cho, R. W. Conners and P. A. Araman, A comparison of rule-based, *k*-nearest neighbor, and neural net classifiers for automated industrial inspection, *ICDMESP-91, Washington* (1991) 202–209.
25. J. H. Friedman and W. Stuetzle, Projection pursuit regression, *J. Amer. Statist. Assoc.* **76** (1981) 817–823.
26. M. Stone, Cross validatory choice and assessment of statistical predictors (with discussion), *J. R. Statist. Soc.* **B36** (1974) 111–147.
27. G. Cybenko, Approximations by superpositions of a sigmoidal function, *Math. Control Signal Syst.* **2** (1989) 303–314.
28. J. Denker, D. Schwartz, B. Wittner, S. Solla, R. Howard, L. Jackel and J. Hopfield. Automatic learning, rule extraction and generalization, *Complex Systems* **1** (1987) 877–922.
29. S. Sjoberg and L. Ljung, *Overtraining, Regularization and Searching for Minimum in Neural Networks*, Technical report, University of Linkoping, 1992.
30. F. W. Dumm, Magnetic particle inspection fundamentals, *Mater. Eval.* **35** (1977) 42–ff.
31. C. A. Gregory, V. L. Holmes and R. J. Roehrs, Approaches to verification and solution of magnetic particle inspection problems, *Mater. Eval.* **30** (1972) 219–ff.
32. G. M. Massa, Finding the optimum conditions for weld testing by magnetic particles, *NDT International* **9** (1976) 16–ff.
33. D. C. Jiles, Review of magnetic methods for nondestructive evaluation (Part 2), *NDT International* **23** (1990) 83–92.
34. Y. F. Cheu, Automatic crack detection with computer vision and pattern recognition of magnetic particle indication, *Mater. Eval.* **42** (1984) 1506–1510.
35. E. Barnard and D. Casasent, Invariance and neural nets, *IEEE Trans. Neural Networks* **7**, 1 (1996) 206–213.
36. M. Fukumi, S. Omotu, F. Takeda and T. Kosaka, Rotation invariant neural pattern recognition system with application to coin recognition, *IEEE Trans. Neural Networks* **3**, 2 (1992) 272–279.

37. Y. LeCun, B. Boser, J. S. Denker, D. Henderson, R. E. Howard, W. Hubbard and L. D. Jackel, Handwritten digit recognition with a backpropagation network, ed. Morgan Kaufmann, *Proc. NIPS-89* (1990) 396–404.
38. R. Leighton, *The Aspirin/Migraines Software Tools*, Technical report, The MITRE Corporation, 1993.
39. J. Denker, Y. L. Cun, P. Simard and B. Victorri, *Tangent Prop – A Formalism for Specifying Selected Invariances in an Adaptive Network*, Technical report, 1992.
40. M. Valle, G. Nateri, D. D. Caviglia, G. M. Bisio and L. Briozzo, An ASIC design for real-time image processing in industrial applications, *IEEE Comp. Soc. Press*, *Proc. ED&TC-95*, Paris, France (1995) 385–390.
41. N. Guglielmi, *A VLSI Architecture for Texture Analysis*, Technical report, University of Bologna, 1993 (In Italian).
42. I. M. Dar, K. E. Newman and G. Vachtsevanos, On-line inspection of surface mount devices using vision and infrared sensors, *IEEE Proc. AUTOTESTCON-95*, Atlanta, GA (1995) 376–384.
43. I. M. Dar, K. E. Newman and G. Vachtsevanos, Bond validation of surface mount components using a combined IR/Vision system, *SME Proc. Applied Machine Vision-94* **2**, Minneapolis, MN (1994).
44. J. Serra, *Image Analysis and Mathematical Morphology* (Academic Press, 1982).

CHAPTER 3

COLLABORATIVE OPTIMIZATION AND KNOWLEDGE SHARING IN PRODUCT DESIGN AND MANUFACTURING

MASATAKA YOSHIMURA

Department of Precision Engineering,
Graduate School of Engineering,
Kyoto University,
Kyoto, 606-8501, Japan
E-mail: yoshimura@prec.kyoto-u.ac.jp

In current and future product design and manufacturing, "collaboration" among different groups, divisions, and/or enterprises is considered one of the most important methodologies. In this chapter, collaboration during product design and manufacturing is discussed from the following two viewpoints: (1) Collaboration in which two decision-makers having different requirements and knowledge cooperatively evolve solutions concerning product design and manufacturing and (2) collaboration in which different groups and/or enterprises having competitive relationships consider collaborative projects. In the above two scenarios, methodologies and practical procedures for realizing superior product design solutions are especially necessary when many factors must be simultaneously evaluated. In this chapter, the principal areas of focus are product design optimization methodologies and practical procedures in which multicriteria optimization techniques are used in concert with human decision making. Collaborative work on projects is usually the most promising strategy for the efficient development of new products. At the present time, people can easily acquire knowledge from widely separated individuals by using computer networks, and cooperative development of projects that use designers, design groups or a number of enterprises, can be efficiently conducted using such networked systems.

Keywords: Collaboration; optimization; knowledge sharing; product design, product manufacturing.

1. Introduction

Individual designers have a limited range of knowledge but the scope of such knowledge can be enlarged by sharing information, either among separate groups of designers or among individual designers within a group. In today's engineering circumstances, where innovation often plays a key role in designing superior products, the acquisition of new knowledge plays an important factor in product design. The gathering and preparation of such knowledge by a single designer, or a small group of designers, is often difficult, time consuming and costly. Sharing information among separate groups of designers, however, can extend the boundaries of existing knowledge. Collaborative work on projects is usually the most promising strategy for the efficient development of new products. Presently, people can easily acquire knowledge from widely separated individuals by using computer networks, and the cooperative development of projects that use designers, design groups or a number of enterprises can be efficiently conducted using such networked systems.

Practical product design is often an extremely complicated process due to the interrelationship of decision variables and various requirements for the product. With this in mind, concurrent engineering[1–4] offers an effective and powerful methodology for realizing the most satisfying product designs possible, from an integrated and global viewpoint.

In current and future product design and manufacturing, "collaboration" among different groups, divisions, and/or enterprises is considered one of the most important methodologies. In this chapter, collaboration during product design and manufacturing is discussed from the following two viewpoints:

(1) Collaboration in which two decision-makers having different requirements and knowledge cooperatively work out solutions pertaining to product design and manufacturing.
(2) Collaboration in which different groups and/or enterprises having competitive relationships consider collaborative projects.

In the above two scenarios, methodologies and practical procedures for realizing superior product design solutions are especially necessary when many factors are to be simultaneously evaluated. In this chapter, the principal areas of focus are product design optimization methodologies and practical procedures in which multicriteria optimization techniques are used together with human decision making.

In scenario (1), a variety of requirements must be satisfied for a product design to be successful. Product designers, process planners and manufacturers each have their own set of requirements, above the fundamental need to maximize user satisfaction and product versatility. Decision-making procedures for simultaneous product design are completely different from those employed in conventional decision making. A methodology for concurrent decision making among decision makers having different requirements is necessary for obtaining better product designs from a system-wide viewpoint. First, a function expressing the satisfaction level for the specific product design is formulated for each of the decision makers. Next, an integrated higher ranking decision-making utility function incorporating the satisfaction level functions is formulated with the cooperative consultation of the decision makers. Then, the most preferable design solution is selected from candidate design solutions using the integrated utility function. The cooperative decision making process makes use of a computer during the calculation of candidate solutions and the graphical presentation of data to the decision makers.

Scenario (2) focuses especially on cooperative new product development projects among different enterprises, each of which must maintain a position that guards its own interests. In order to conduct such collaborative work effectively, it is necessary to quantitatively evaluate the benefits of knowledge sharing in a given collaborative project, yet such determinations have not often been attempted to date. Such numerical evaluation methods enable the viability of the cooperative project to be accurately determined. The possibilities for generating new design solutions that cannot be obtained by an isolated designer, or by the simple addition of knowledge from other independent designers, are first discussed with reference to knowledge sharing in cooperative projects. Possibilities in creating new knowledge come into existence when designers in different fields share what they know. But cooperative work is feasible only when each partner can mutually benefit from sharing his or her knowledge. In the latter part of this chapter, a numerical measure expressing the viability of such cooperative work is explained and practical procedures for determining the viability of cooperative work with designers are described. Using these methods, synergy effects can be quantitatively evaluated by viewing changes in the Pareto optimum solution sets.

2. Background of Concurrent Optimization and Collaboration

In this chapter, the following two types of collaborative engineering techniques are discussed:

(1) Collaborative decision making between decision makers having different requirements.
(2) Knowledge sharing among different people, groups or enterprises.

2.1. *The need for simultaneous decision making, and a new decision making strategy*

Figure 1(a) shows the structural flow for conventional sequential decision making, from top to bottom, for a case having three decision making divisions. In this one-way sequential process, factors determined in the upper stage become constraints in the lower (i.e. later) stages of decision making. Figure 1(b) shows a concurrent decision making structure in which each division is placed at the same level and concurrent decision making is conducted by all divisions, incorporating horizontally wider, unrestricted viewpoints.

The information relating to a product's design should be concurrently utilized at the highest level of the decision making processes, where most important product design factors are considered. To do this, methodologies are needed that will allow the simultaneous evaluation of a range of factors within a given arena of the overall product design process.

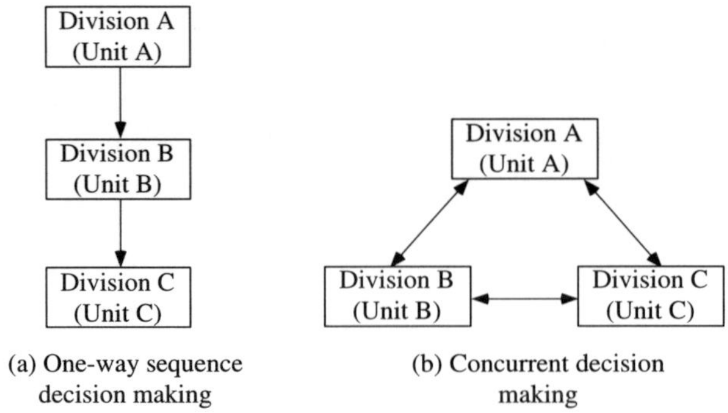

Fig. 1. Types of decision making structures.

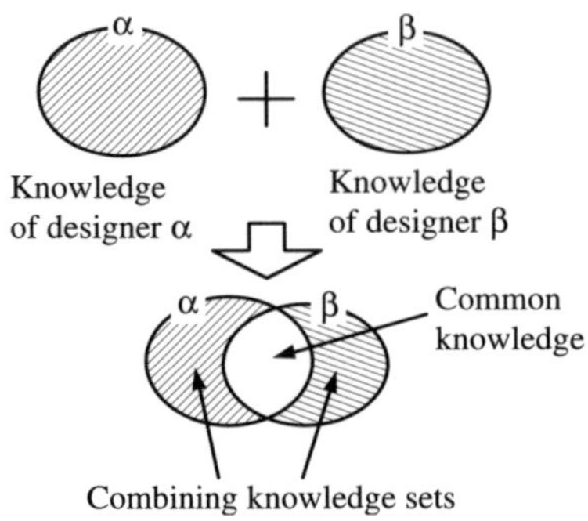

Fig. 2. The concept of knowledge sharing.

2.2. *The need for knowledge sharing*

Each designer has a limited range of knowledge concerning a given product that is being designed. The easiest and quickest methods of obtaining new or wider knowledge, i.e. broadening the range of knowledge for a given designer, is to make use of other designers' knowledge. Figure 2 shows the concept of knowledge sharing among designers α and β. Knowledge that one designer lacks may be supplied by another. Possibilities in creating new knowledge come into existence when designers in different fields share what they know. The effective use of such external knowledge and information is often of key importance for the development of new products.

3. Evaluation in Concurrent Optimization

To clarify the relationships between conflicting characteristics that appear during the design process, it is effective to formulate a multiobjective design optimization problem in which the characteristics are simultaneously evaluated on the same stage.[5,6] Multiobjective optimization methods are used to concurrently evaluate characteristics and the optimum solutions are determined from a global viewpoint.

When optimizing the design of machine products, higher operational accuracy, shorter operation time, smaller operating cost, and smaller manufacturing cost are all preferable characteristics. Therefore, optimal solutions should be selected from the points on the line of the Pareto optimum solution set of a multiobjective optimization problem which ideally has all, or at least several, of the following objectives: maximization of operational accuracy; minimization of operation time; minimization of operating cost; and minimization of manufacturing cost.

Within product design divisions, product performances are commonly evaluated according to how well the required product functions. Process designs are conducted in manufacturing divisions, where the most practical methods for manufacturing the designed products are determined, and the manufacturing costs evaluated. Thus, in the overall decision making of product design and manufacturing, product performance and manufacturing cost are the principal evaluative characteristics.

Product designers, by their nature, seek higher product performance, while process planners seek to lower product manufacturing cost, two characteristics which are in mutual competition. Figure 3 shows the relationship between product performance and product manufacturing cost.[7] The shaded part corresponds to the feasible design region when using present technologies, knowledge and/or theories. The designers are searching for design solutions in the direction of the big arrow, as shown in Fig. 3. The solid line PQ corresponds to the Pareto optimum design solutions of a multiobjective optimization problem, representing specific feasible design solutions that will yield an improvement in one objective without causing a degradation in the other objectives.[8]

The present problem has two objectives: (1) maximization of product performance and (2) minimization of product manufacturing cost. The solutions are a set of design points where both further improvement of product performance, and

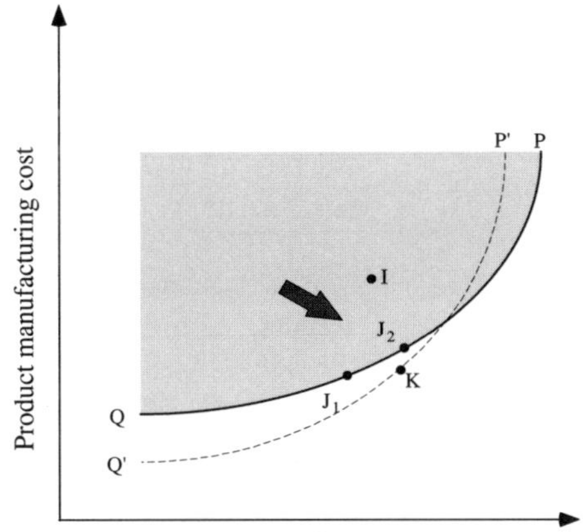

Fig. 3. Relationship between product performance and product manufacturing cost.

further reduction of product manufacturing cost, are impossible. The designers will ultimately search for a design solution lying on the Pareto optimum line.

A Pareto optimum solution set, such as that shown in Fig. 3, indicates the candidate solutions among which the most suitable solution or the optimum solution will be selected.

In the process of decision making while designing products, a high level evaluative criterion is used which integrates and expresses the overall satisfaction level for the product design solution. A design solution having a maximum value of a specific satisfaction level (for example, that of the manufacturers of the product, the users of the product, or the degree to which market needs are satisfied) is selected as the optimum one.

In the context of such concurrent or simultaneous decision making procedures, the most important decisions should be conducted by human beings. However, the use of computers can often help decision makers make the best possible decisions.

4. Advantages of Collaborative Optimization and Design Collaboration

4.1. *Advantages of concurrent optimization*

Product design divisions and manufacturing divisions are fundamentally related during the development of the product's design, and when they both arrive at a single solution on the Pareto optimum solution line (as shown by solid line PQ), the design problem is solved. When the requirements of different divisions conflict with each other, say in terms of performance or cost, the requirements of the various design divisions should be concurrently evaluated on a level stage, and not according to a decision making hierarchy.[9]

Figure 4 shows the flow sequence of a conventional one-way design process. In the product design division, the designs of a product are first determined after defining the specifications of the product according to market needs. Then these detailed designs are transferred to the manufacturing division, where detailed manufacturing

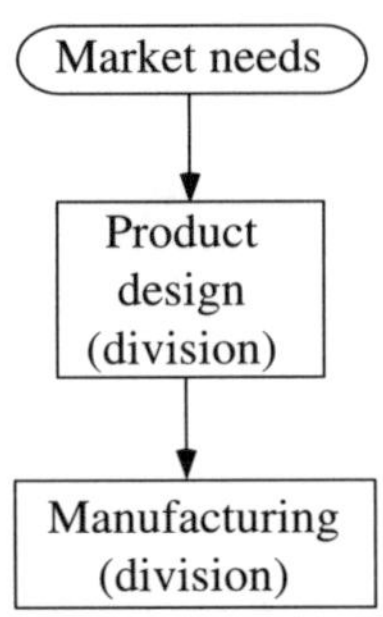

Fig. 4. Conceptual diagram of a conventional one-way sequential design.

methods are determined accordingly, so that a product that satisfies the product specifications can be manufactured.

The following problems occur with this kind of conventional, one-way process, where products are developed linearly from the design stage to the manufacturing stage:

(1) Since the product design division lacks certain knowledge and information concerning the manufacturing process, the design solution does not exist on the Pareto optimum solution line of the objective functions space, as illustrated at Point I in Fig. 3.
(2) Even if the design solution is located on the Pareto optimum solution set line, such as point J_1 in Fig. 3, a solution having a higher level of satisfaction may exist at another point (for example, at point J_2) which is also on the Pareto optimum solution set line.
(3) A design solution at point K on a more preferable Pareto optimum solution set line, such as shown by a broken line $P'Q'$ in Fig. 3, exists when other materials and/or manufacturing methods are used.

The word "concurrent" has a dual meaning. It is defined as an agreement of opinion, and it also means working or operating together at the same time. In concurrent designs, the limitations of one-way, sequential decision making are transcended, as both product design and manufacturing decisions are simultaneously and cooperatively conducted, as shown in Fig. 5.

In concurrent designs, the evaluation of the manufacturing cost is resolutely conducted at the design stage, so that an accurate estimate of the manufacturing cost can be reasonably acquired. Furthermore, the specific intentions of designers and manufacturers can be mutually understood and agreed upon, preventing costly revisions of designs and manufacturing processes due to fundamental discrepancies. Thus, the lead time of the products can be reduced, which is an increasingly important consideration. From the foregoing discussion, it should be clear that concurrent design procedures can yield products that are more competitive in the marketplace than those whose designs were developed through one-way, sequential design procedures.

Usually, product performance and manufacturing cost requirements are in direct conflict. However, concurrent decision making methods allow these opposing evaluative factors to be simultaneously considered in both the product design and the manufacturing divisions over time, which aids the solution of conflicts and the production of an optimized product.

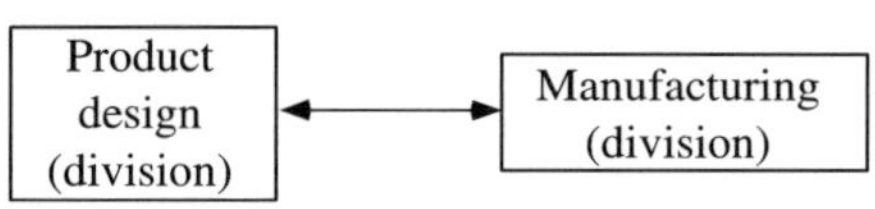

Fig. 5. Uni-level placement of product design and manufacturing, and their concurrent processing.

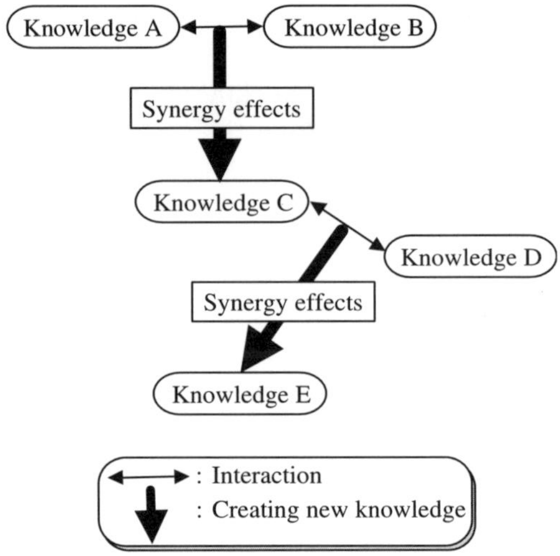

Fig. 6. The concept of synergy effects.

4.2. *Advantages of knowledge sharing and synergy effects*

Synergy means the cooperative action of two or more organizations or groups, usually under mutually beneficial circumstances. The concept of synergy effects is illustrated in Fig. 6. Synergy effects result from the interaction of knowledge A and B, producing new knowledge C. After that, knowledge C may "encounter" knowledge D as the designers work together, and knowledge E is created, due to the interaction of knowledge C and D.

5. Collaborative Decision Making in Product Design and Manufacturing

5.1. *Fundamental procedures*

When there is more than one decision maker, each having individual desires, cooperative decision making is necessary.

When considering conflicting requirements of design and manufacturing divisions simultaneously, manufacturing factors are placed on the same level as those of product design, as shown in Fig. 5. Decision making items relating to these two divisions, product design and manufacturing (which consists of both process design and practical manufacturing) are concurrently being processed. In order to realize a concurrent design, product requirements from the standpoint of both the product design division and the manufacturing division must be clarified first. Thereafter, a convincing solution can then be obtained, based on a reasonable method.

Usually, the realization of better product performance results in higher product manufacturing cost, while the lowering of product manufacturing cost results in lower product performance. Thus, the requirements for acquiring better product performances while lowering manufacturing cost are naturally in conflict. In order to coordinate these requirements, an opinion or criterion which comes from a separate viewpoint, higher ranked than these two themselves, is necessary. Here, the upper ranking decision making function is formulated as such a criterion.

The product design and manufacturing divisions each have their own satisfaction levels for design proposals which take into account the product's environment. First, a single-attribute satisfaction function is formulated from the standpoint of each of the above product's two design divisions. Then, a higher ranking decision making function, and a two-attribute satisfaction function, are defined, using utility analysis procedures.[10]

The single-attribute satisfaction function is one where a satisfaction level s, having a value from 0 to 1 on the vertical axis, is defined for the value of ε. ε expresses the difference between the product characteristic value and the required value (the goal value) on the horizontal axis,[11] as shown in Fig. 7.

Here, the following function is used as the fundamental form of the satisfaction function:

$$s(\varepsilon) = \frac{1}{\pi} \tan^{-1}\{a(\varepsilon + b)\} + 0.5. \tag{1}$$

The form of the satisfaction function can be modified by changing parameters a and b of Eq. (1). Smaller values of parameter a give a gentler rise to the function, while a larger values of parameter a make the rise sharper. Parameter b corresponds to a value of ε having a satisfaction level of 0.5. That is, parameter b defines the magnitude of ε in a compromised result having a 50% satisfaction level. Here, in order to facilitate interactive decision making concerning the form of the satisfaction function, parameters b and s_0 of the satisfaction level at $\varepsilon = 0$ as shown in Fig. 7

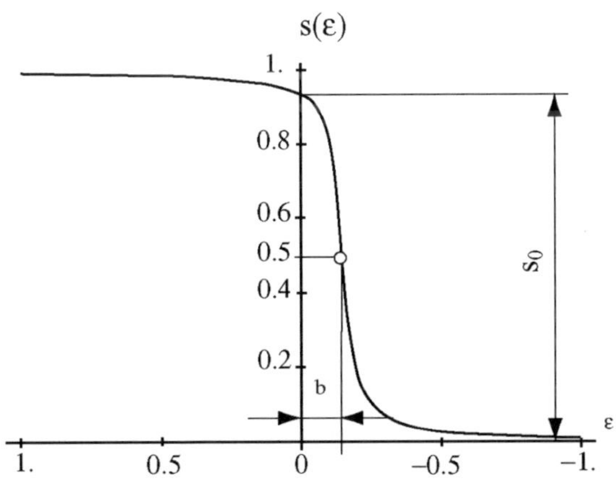

Fig. 7. Features of a single attribute satisfaction function.

are used. Parameter a in Eq. (1) is obtained using b and s_0 as follows:

$$a = \frac{\tan\{\pi(s_0 - 0.5)\}}{b}. \tag{2}$$

First, the decision maker provides the initial values of s_0 and b while considering the requirements of his or her circumstances. The form of the satisfaction function, such as that shown in Fig. 7, is then displayed. If the decision maker is not satisfied with the form of the function, he can modify the values of b and s_0. These procedures are repeated until the decision maker is satisfied with the form of the satisfaction function.

Figure 8(a) shows an example of the form of the single-attribute satisfaction function $s_1(y)$ which indicates a product design decision maker's satisfaction level as the product performance y varies. Figure 8(b) shows an example of the form of

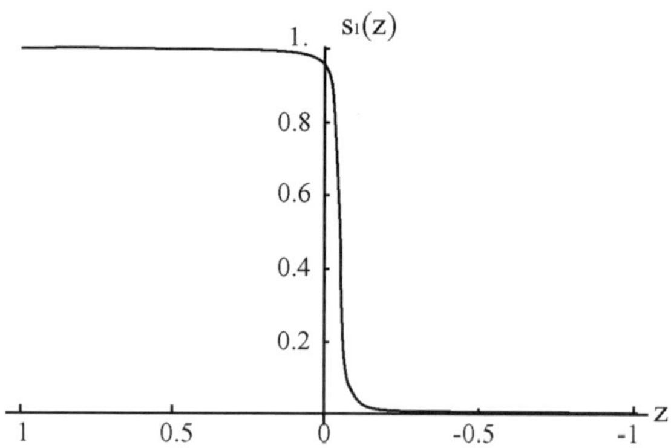

(a) Product designer's satisfaction function for product performance.

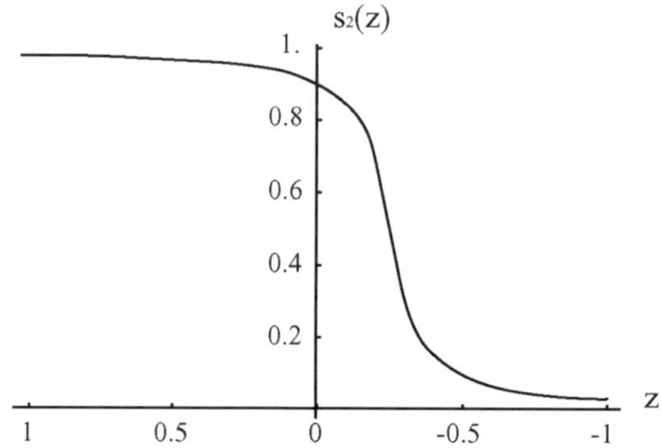

(b) Process planner's satisfaction function for product manufacturing cost.

Fig. 8. Examples of satisfaction function.

the single-attribute satisfaction function $s_2(z)$ which indicates a decision maker's satisfaction level as the manufacturing cost z is varied.

The two-attribute satisfaction function $s(y, z)$[12] is obtained using the single-attribute satisfaction function for each attribute of y (product performance) and attribute of z (product manufacturing cost) as follows:

$$s(y, z) = k_y s_1(y) + k_z s_2(z) + k_{yz} s_1(y) s_2(z), \tag{3}$$

where $s_1(y)$ is the product designer's single-attribute satisfaction function concerning product performance and $s_2(z)$ is the process planner's single attribute satisfaction function concerning product manufacturing cost.

The coefficients k_y, k_z and k_{yz} in Eq. (3) are determined by methods similar to those used in the two-attribute utility analysis, as explained below.

Consider a situation where there are two kinds of products, D_1 and D_2. Product D_1 has the best product performance y^*, but the highest manufacturing cost z^0 within the decision making area is shown by point A in Fig. 9. When product D_2 has the lowest possible manufacturing cost z^* within the decision making area, the product performance required of product D_2 in order to offer a satisfaction level equivalent to that of product D_1 is y'. y' is the product performance at point C in Fig. 9.

The following relation, as expressed in Eq. (4), is established by applying to Eq. (3) the fact that product D_2 offers the same level of satisfaction as product D_1:

$$k_y = s_1(y') + k_z\{1 - s_1(y')\}. \tag{4}$$

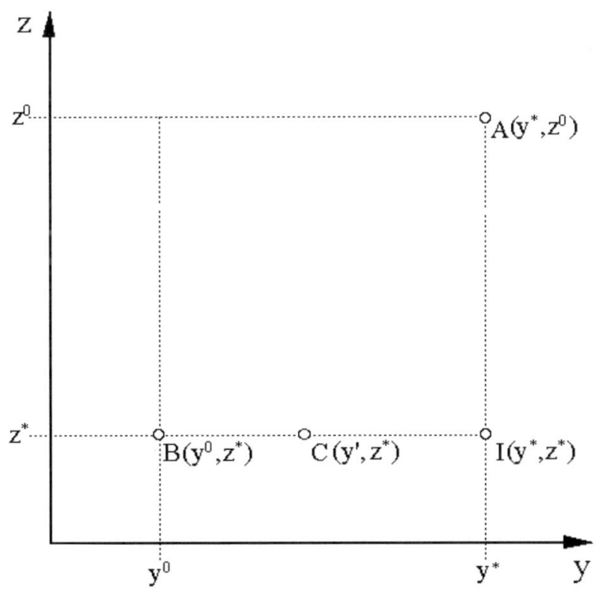

Fig. 9. Explanation diagram for the determination of y'.

The value of y' at point C is not decided by product designers alone, but it is determined during the cooperative consultation of product designers and process planners. Since this value has a direct influence on coefficients k_y, k_z and k_{yz} of the two-attribute satisfaction function, careful consideration is necessary. This point represents a compromise design solution, where the satisfaction levels of the two divisions, design and manufacturing, are equal.

With the satisfaction level at point I in Fig. 9 being 100% (that is, $s(y^*, z^*) = 1$), the coefficients k_y, k_z and k_{yz} have the following relation at that point:

$$k_y + k_z + k_{yz} = 1. \tag{5}$$

In utility analysis, it is usual for coefficient k_z to be determined based on a concept of certainty equivalence using lotteries,[12] but this is not the case here. The coefficient is here determined as follows:

Using parameters b_y and b_z of single-attribute satisfaction functions, the following relation which gives a two-attribute satisfaction level of 50% is defined as:

$$s(b_y, b_z) = 0.5, \tag{6}$$

where b_y indicates a product performance value which gives a 50% satisfaction level for product designers, while b_z indicates a product cost which gives a 50% satisfaction level for process planners. From Eqs. (4), (5) and (6) and the relations of $s_1(b_y) = 0.5$ and $s_2(b_z) = 0.5$, the coefficients k_y, k_z and k_{yz} can be determined and the two-attribute satisfaction function $s(y, z)$ can be obtained.

5.2. *Example of concurrent design*

Now, the design of a cantilever model having stair-step shapes, as shown in Fig. 10, is considered. Here, the candidate materials are alloys of aluminum or magnesium. Casting and forging are considered as the candidate manufacturing methods.

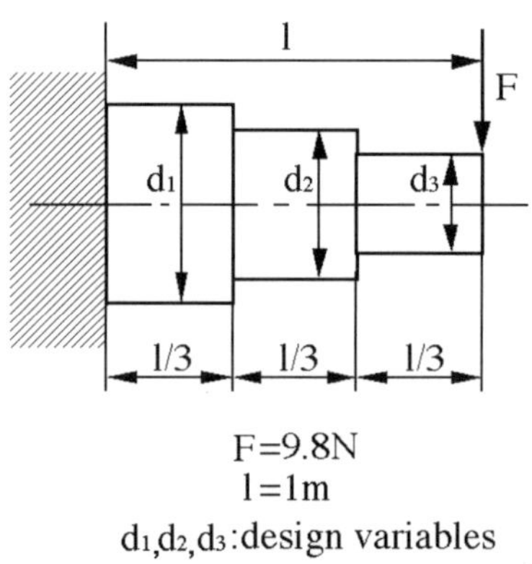

Fig. 10. A cantilever model having stair-step shapes.

The design solution was determined by having both the displacement δ at the free end of the model and the manufacturing cost M_{cost} as small as possible, under the constraint of the product weight W. Here, the upper limit of the product weight W is 490.3 N (50 kgf).

The design optimization problem was formulated as a multi-objective optimization problem as follows:

$$\begin{aligned} \psi = \psi[\delta, M_{\text{cost}}] &\rightarrow \text{minimize} \\ \text{subject to } W &\leq 490.3 \text{ [N]}. \end{aligned} \tag{7}$$

A Pareto optimum solution set is obtained for the two objectives namely: (1) Maximizing the product performance — which is the principal evaluative characteristic of the product design division, and (2) Minimizing the manufacturing cost — the principal evaluative characteristic of the manufacturing division. The Pareto optimum solution set curves obtained by solving the multi-objective optimization problem are shown in Fig. 11. The Pareto optimum solution line shows the relationship between the product performance and the manufacturing cost for the candidate design solutions from which the optimum solution will be finally selected.

In the next step, the two-attribute satisfaction function for the product design division and the manufacturing division was formulated for the design model.

Representatives of the product design division and the manufacturing division concurrently selected a compromise point, analogous to point C in Fig. 9, by engaging in cooperative decision making. In this example, in terms of displacement at

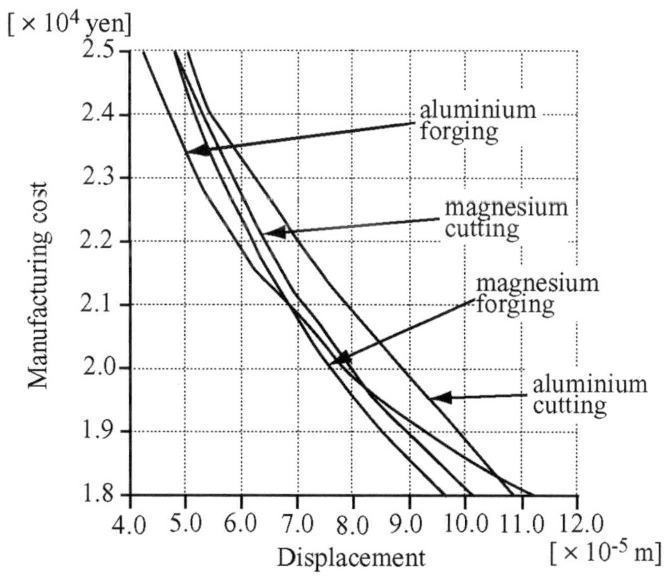

Fig. 11. Pareto optimum solution curves for the stair-stepped cantilever model.

the free end of the model, the value of the product performance at point y' on the y-axis was 6.86×10^{-5} [m].

At this point, the value of the single attribute satisfaction function, $s_1(y')$, was 0.493.

Then, the coefficients k_y, k_z and k_{yz} were obtained, yielding the following values:

$$k_y = 0.6637, \quad k_z = 0.3362, \quad k_{yz} = 1.0 \times 10^{-4}.$$

Since coefficient k_{yz} is negligibly small, k_{yz} was set to 0. Then, the two-attribute satisfaction function was obtained as follows:

$$s(y, z) = 0.6637 s_1(y) + 0.3362 s_2(z). \tag{8}$$

The features of this function are shown in Fig. 12.

The satisfaction levels on the Pareto optimum solution curves shown in Fig. 11, which were obtained using the two attribute satisfaction function, are shown in Fig. 13. The design solution having the maximum satisfaction level, selected as the most preferable design, was as follows:

Material: aluminum alloy.
Manufacturing method: forging.
Displacement at the free end of the product model: 6.4021×10^{-5} m.
Manufacturing cost: 21,300 ¥.

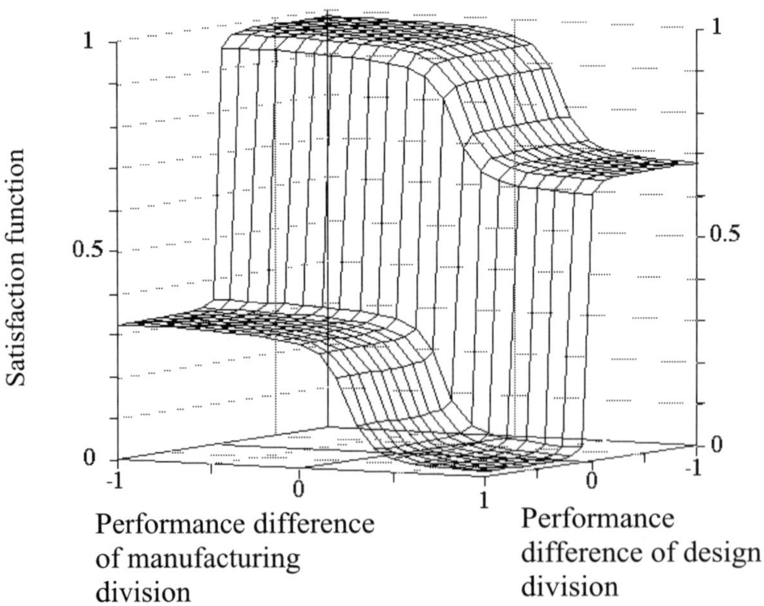

Fig. 12. Two-attribute satisfaction function curves of product design and manufacturing.

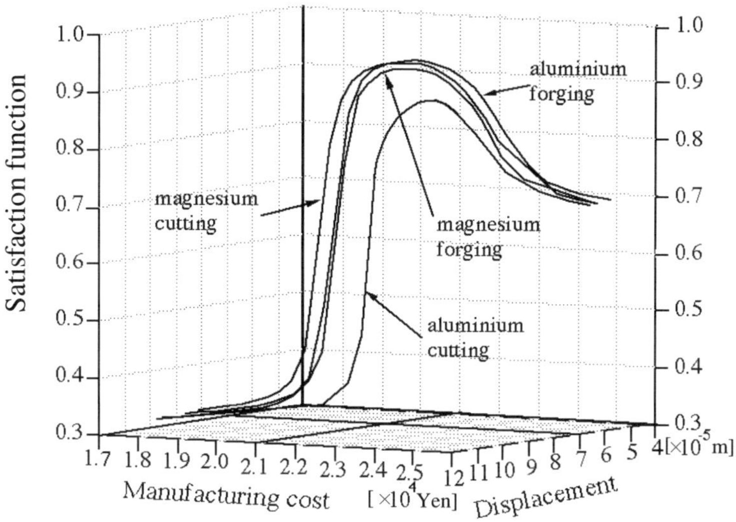

Fig. 13. Satisfaction levels on the Pareto optimum solution curves of the stair-stepped cantilever model.

In a one-way sequential design process, the kind of material used and the manufacturing method are selected at the outset, from a simple case database. In concurrent design, however, the design solution is arrived at by examining all possible combinations of the kinds of materials and manufacturing methods, during the search for an optimum solution. Therefore, in a concurrent design, design solutions inaccessible in one-way sequential design methods can be readily obtained.

6. Knowledge Sharing in Collaborative Projects

6.1. *Methodologies for evaluating the effectiveness of knowledge sharing*

When a designer cooperatively develops a new product with other designers, he usually expects to obtain additional knowledge from the other designers during the normal process of information exchange occurring in such projects. If none of the designers can obtain additional knowledge or if only one designer can acquire such new knowledge, the work cannot be known as cooperative. Truly cooperative work is established when a group of designers can share knowledge among themselves. That is, cooperative work is feasible only when each partner can mutually benefit from sharing his or her knowledge.

Here, numerical measures for evaluating benefit levels of cooperative projects are explained, and methodologies for evaluating whether or not cooperative work is viable are given.[13] Then, the design solutions obtained by knowledge sharing are compared with those arrived at by isolated designers.

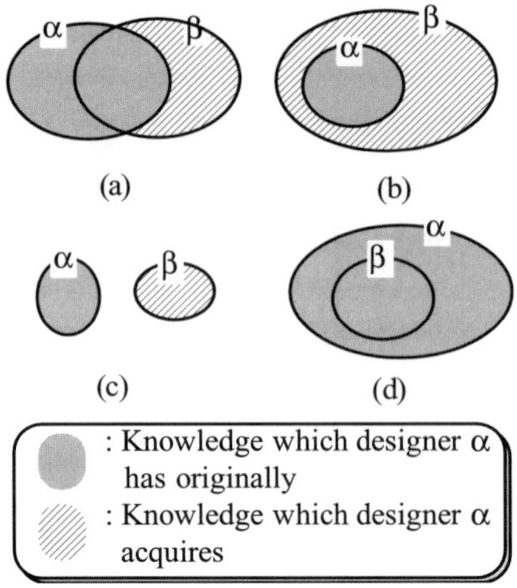

Fig. 14. Pattern of knowledge sharing among designers α and β from α's point of view.

Here, knowledge sharing among only two designers is mainly considered. The methodologies constructed for two designers can easily be expanded and applied to cases where the number of designers is greater.

6.2. *Patterns of knowledge sharing*

The patterns of knowledge sharing among two designers α and β can be categorized from designer α's point of view as follows:

(1) Designer α at the outset possesses some of the knowledge that designer β has. Designer α can acquire the remaining portion of β's knowledge by knowledge sharing, as shown in Fig. 14(a).
(2) Designer β at the outset possesses all of the knowledge that designer α has, as shown in Fig. 14(b).
(3) Designers α and β have no common knowledge, as shown in Fig. 14(c).
(4) Designer α at the outset has all of the knowledge that designer β has, as shown in Fig. 14(d).

For designer α to obtain benefits by knowledge sharing, acquiring knowledge that he does not have is a necessity. Thus, in pattern (d) of the knowledge sharing figures, he obtains no benefit. Similarly, in pattern (b), designer β does not obtain any benefits. In patterns (a) and (c), however, both designers have the possibility of mutually obtaining benefits, and synergy effects due to knowledge sharing can be expected.

6.3. *Practical procedures to determine the viability of cooperative work*

A numerical measure for determining the viability of such cooperative work is now explained. Practical procedures for determining the viability of cooperative work with designers α and β use this measure as follows:

(1) Selection of items to be evaluated in product design:
The items to be evaluated in the product design are denoted by I_i ($i = 1, 2, \cdots, N$), where N is the total number of items.
(2) Judging whether or not each designer can acquire new knowledge concerning each item:
Whether or not each of the two designers α and β can acquire new knowledge when the knowledge of item i shared between them is set up as follows:

$$\begin{aligned}
\omega_{\alpha i} &= 1 \quad \text{when designer } \alpha \text{ acquires new knowledge.}\\
\omega_{\alpha i} &= 0 \quad \text{when designer } \alpha \text{ does not acquire new knowledge.}\\
\omega_{\beta i} &= 1 \quad \text{when designer } \beta \text{ acquires new knowledge.}\\
\omega_{\beta i} &= 0 \quad \text{when designer } \beta \text{ does not acquire new knowledge.}
\end{aligned}$$

(3) Defining the importance levels of items:
When new knowledge is obtained, acquisition of the knowledge is meaningless for the project if the knowledge is not important for the given designer. Designers α and β define the importance level $s_{\alpha i}$ and $s_{\beta i}$ of item i, respectively, using the pairwise comparison matrix of the AHP (Analytic Hierarchy Process) method.[14]
(4) Defining benefit levels of knowledge sharing:
Designer α's benefit level, S_α, obtained by sharing knowledge of all of the items is calculated as follows:

$$S_\alpha = \sum_i \omega_{\alpha i} s_{\alpha i}. \tag{9}$$

Similarly, designer β's benefit level S_β is calculated as:

$$S_\beta = \sum_i \omega_{\beta i} s_{\beta i}. \tag{10}$$

(5) Determination of the viability of the cooperative project:
The viability of the cooperative project is determined by using the procedures as outlined in Fig. 15. First of all, if either designer α or β can acquire new and important knowledge, the cooperative project is considered to be viable. Next, when both designers' benefit levels under knowledge sharing are rather high, cooperative work is judged to be viable according to the following evaluation.

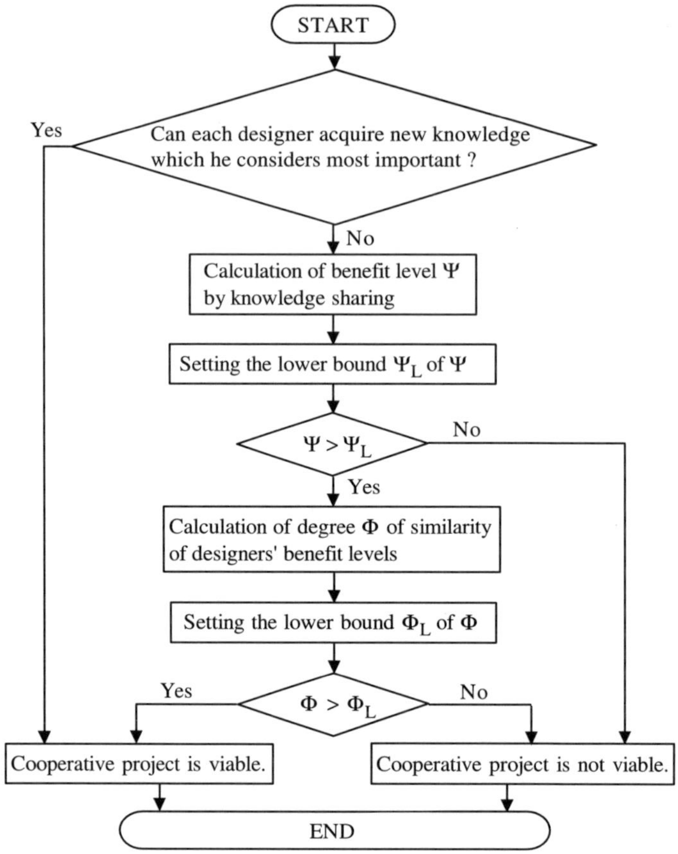

Fig. 15. The flowchart to determine the cooperative project's viability.

The product of designers α's benefit level S_α and β's benefit level S_β is denoted by Ψ:

$$\Psi = S_\alpha S_\beta. \tag{11}$$

With Ψ_L as the lower bound of Ψ, when $\Psi \leq \Psi_L$, the cooperative project is judged to be inviable. The default value of Ψ_L is 0.25, the value when each designer's benefit level is 0.5. The value of Ψ_L can be adjusted, depending on the needs of both designers.

When $\Psi > \Psi_L$, a further evaluation is conducted. If neither designers can acquire useful knowledge, or if their benefit levels differ considerably, the cooperative project is not viable. The ratio of the two designers' benefit levels is evaluated as follows:

$$\Phi = \frac{\min\{S_\alpha, S_\beta\}}{\max\{S_\alpha, S_\beta\}}, \tag{12}$$

where Φ is $0 \leq \Phi \leq 1$. The cooperative work is viable when $\Phi > \Phi_L$, while it is not viable if $\Phi \leq \Phi_L$. The default value of Φ_L here is 0.5, i.e. the benefit level of the designer having a higher benefit level is twice as high as the benefit level of the other designer. Φ_L can also be altered at the request of either designer.

6.4. *Consideration of synergy effects via knowledge sharing*

In usual product design optimization formulations, there are at least two principal product characteristics included in objective functions or constraint functions that have conflicting relationships with each other. Now, a product design optimization problem having two principal product characteristics P_1 and P_2 is considered, where smaller values of both P_1 and P_2 are preferable.

Design optimization formulations differ depending on the requirements which the products have to meet, and these are often determined by the product's working environments. Generally, there are two formulations for design optimization, which reflect the dual objective and constraint functions of the product's characteristics. Often, one of the two functions is given priority, and here, P_1 is given the highest priority.

Formulation A
$P_1 \to$ minimize
subject to $P_2 \leq P_2^U$
and other constraints.
Formulation B
$P_2 \to$ minimize
subject to $P_1 \leq P_1^U$
and other constraints.

Formulation A corresponds to a case in which the best feasible value of the most important characteristic (P_1) is obtained. Formulation B corresponds to a case in which P_1 is subjected to severe constraints while the other principal characteristic is minimized.

The combined form of Formulations A and B is expressed as a multiobjective optimization problem in which both P_1 and P_2 are included in the objective functions as follows:

$$\psi[P_1(x), P_2(x)] \to \text{minimize}$$
$$\text{subject to constraints.}$$

The optimum solutions of Formulations A and B are included in the Pareto optimum solution set of the foregoing multi-objective optimization problem. Hence, the line of the Pareto optimum solution set broadly expresses the features of the optimum solutions of the design problems being regarded. Figure 16 shows the two-objective function space of P_1 and P_2. The shaded areas indicate feasible design regions.

If designer α considers product characteristic P_2 as the most important and if he already possesses good experience and knowledge for realizing these requirements,

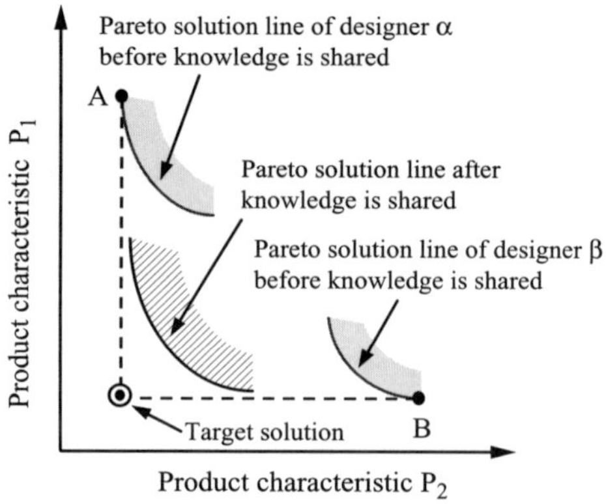

Fig. 16. Comparison of solutions of two designers before and after knowledge sharing.

he can obtain the design solution from among the Pareto optimum solution set line as shown on the upper left side of Fig. 16. On the other hand, if designer β considers product characteristic P_1 to be the most important and has good experience and knowledge for realizing such requirements, he can obtain the design solution from among the Pareto optimum solution set line as shown on the lower right side of Fig. 16.

In the optimization formulation after knowledge sharing occurs, the feasible design region defined by constraints becomes broader than the corresponding region in the optimization before knowledge sharing. The feasible design regions which are defined by design constraints are enlarged by combining the two designers' knowledge. When discrete variables are denoted by X_j $(j = 1, 2, \cdots, K)$, where K is the total number of discrete variables, which are feasible for use by designers α and β are denoted by $X_{j\alpha}$ and $X_{j\beta}$, respectively. After knowledge sharing among designers α and β, feasible variables are denoted by $X_{j\alpha} \cup X_{j\beta}$. The constraints after knowledge sharing are as follows:

$$x_j \in (X_{j\alpha} \cup X_{j\beta}), \quad j = 1, 2, \cdots, K.$$

In the decision making method explained here, synergy effects can be quantitatively evaluated by viewing changes in the Pareto optimum solution sets before and after knowledge sharing, as shown in Fig. 16. The target solution point is an ideal point having two characteristic values, each of which is the best feasible characteristic value realized by either designer α or β. Synergy effects are realized when a solution nearer to the target solution point is obtained due to knowledge sharing.

6.5. *Example*

6.5.1. *Problem description*

Applied examples of a project to design industrial robots are given to illustrate these synergy effects.

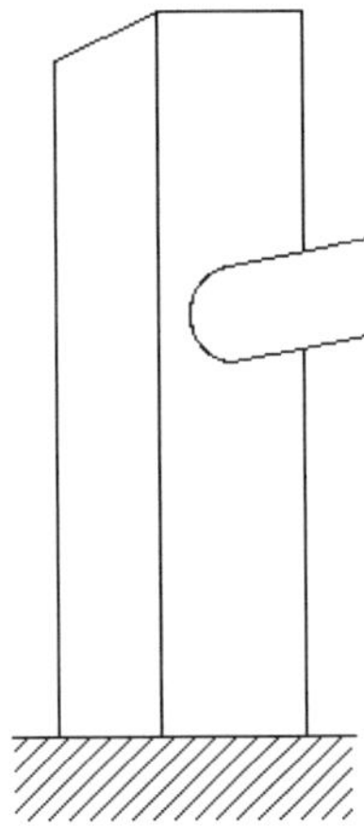

Fig. 17. The overview of a horizontally articulated robot.

The product model, developed in a cooperative project having two designers α and β, is a horizontally articulated robot, as shown in Fig. 17. The areas of knowledge required for designing the product are materials (item I_1), arm shapes (item I_2), and motors (item I_3). The characteristics to be evaluated are the total mass W of the structure, the maximum displacement δ at the end-effector point, and the operation time T.

Related to the cooperative project, each designer, α and β, possesses the following kinds of knowledge in his or her working circumstances at the outset:

Designer α has thorough experience in developing industrial robots used in working circumstances that require high operational efficiency. He has the usage experience with high-speed motors and he knows how to realize lightweight designs of moving arms that exploit their advantages. However, his knowledge concerning materials is poor.

Designer β has adequate experience in developing industrial robots used in working circumstances requiring high operational accuracy. His knowledge and usage experience concerning materials have helped his designs realize higher accuracy. However his knowledge of motors for high speed movement of arms is poor.

In the above engineering knowledge situation, a project to develop a product whose performance achieves both high operational efficiency and high operational accuracy is considered in which each of the two designers offers knowledge to the other and shares such knowledge mutually.

Here, two objective functions, P_1 and P_2, in the multiobjective optimization formulation in Sec. 6.4, correspond to the maximum displacement δ at the end-effector point, and the operation time T, respectively. The total mass W of the structure is constrained as $W \leq 14\,\text{kg}$.

Designers α's and β's pair-comparison matrix of items which are evaluated during the product design, and the importance levels for these items, are shown in Tables 1 and 2, respectively.

The knowledge of each item (I_1: materials, I_2: cross-sectional shapes, and I_3: motors) that each designer α and β has at the outset is as follows:

$I_{1\alpha}$ = {mild steel, aluminum alloy}
$I_{1\beta}$ = {mild steel, aluminum alloy, FRP}
$I_{2\alpha}$ = {hollow cross-section}
$I_{2\beta}$ = {solid cross-section}
$I_{3\alpha}$ = {Motor No. 2, Motor No. 3}
$I_{3\beta}$ = {Motor No. 1, Motor No. 2}.

Figure 18(a) shows a structural model for evaluating the displacement at the end-effector point. M is the mass element of the end-effector, including the object

Table 1. Pair comparison results of designer α.

	I_1	I_2	I_3	Importance Level s_i
I_1 : Material	1	7	5	0.731
I_2 : Arm shape	1/5	1	1/3	0.081
I_3 : Motor	1/5	3	1	0.188

Table 2. Pair comparison results of designer β.

	I_1	I_2	I_3	Importance Level s_i
I_1 : Material	1	3	1/5	0.195
I_2 : Arm shape	1/3	1	1/6	0.088
I_3 : Motor	5	6	1	0.717

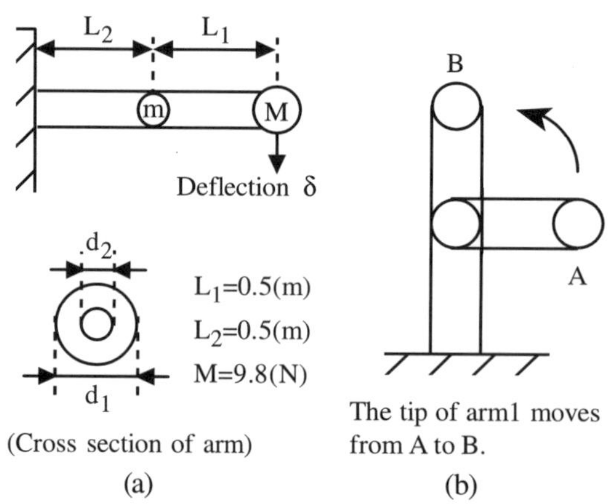

Fig. 18. The model for evaluating the deflection at the end-effector point and operation time.

held, while m is the mass element of a motor. Figure 18(b) shows an illustration explaining the time required for the end-effector point to move from point A to B. Table 3 shows the rated torque, the maximum angular speed, and the weight of the motors. Table 4 shows the Young's moduli and the densities of materials.

Now, the following two cases of cooperative projects having different types of knowledge sharing are considered:

Case 1

This case corresponds to one in which one designer has far more knowledge than the other designer. The relationship of which designer possesses knowledge of which items is shown in Fig. 19. The results of judging whether or not designer α and β can acquire new knowledge for each item are as follows:

$$\text{Designer } \alpha: \ (\omega_{1\alpha}, \omega_{2\alpha}, \omega_{3\alpha}) = (0, 1, 0).$$
$$\text{Designer } \beta: \ (\omega_{1\beta}, \omega_{2\beta}, \omega_{3\beta}) = (1, 1, 1).$$

Case 2

This case corresponds to one in which both designers can obtain new knowledge almost equally. The relationship of which designer possesses which knowledge is shown in Fig. 20. Results judging whether or not designer α and β can obtain new

Table 3. Rated torque, maximum speed, and weight of motors.

Motor No.	Rated Torque τ (N · m)	Maximum Angular Speed $\dot{\theta}_{\max}$ (rad/s)	Motor Weight m (N)
1	35	1.99	9.8
2	51	3.14	14.7
3	65	1.99	19.6

Table 4. Young's moduli and densities of materials.

Material	Young's Modulus E (N/m^2)	Mass Density ρ (kg/m^3)
Mild steel	2.1×10^{11}	7.8×10^3
Aluminum alloy	6.9×10^{10}	2.7×10^3
FRP	5.9×10^{10}	2.0×10^3

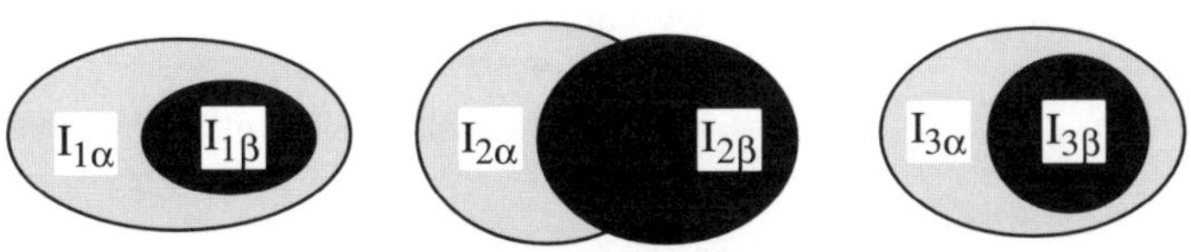

Fig. 19. Knowledge sharing pattern 1 of two designers (Case 1).

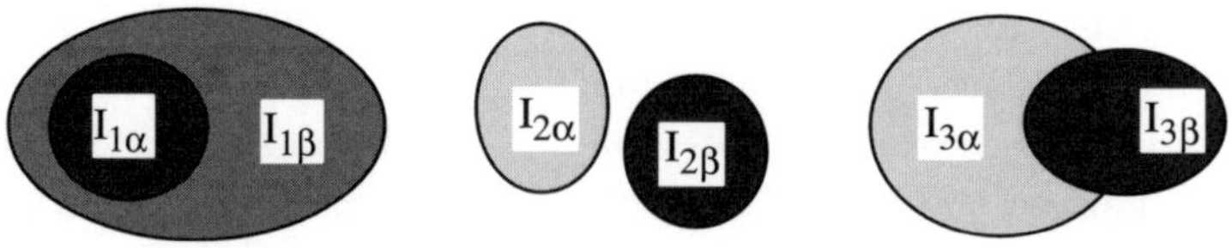

Fig. 20. Knowledge sharing pattern 2 of two designers (Case 2).

knowledge for each item are as follows:

$$\text{Designer } \alpha\colon\ (\omega_{1\alpha}, \omega_{2\alpha}, \omega_{3\alpha}) = (1, 1, 1).$$
$$\text{Designer } \beta\colon\ (\omega_{1\beta}, \omega_{2\beta}, \omega_{3\beta}) = (0, 1, 1).$$

6.5.2. *Decision of the viability of cooperative work*

Case 1

The benefit levels of designers α and β under knowledge sharing were calculated using Eqs. (9) and (10) as follows:

$$S_\alpha = 0.081 \text{ and } S_\beta = 1.$$

Designer β can acquire the knowledge of item I_3 to which he gives the highest importance level, while designer α cannot acquire the knowledge of item I_1 to which he gives the highest importance level (because designer β doesn't have it). Then, Ψ was calculated using Eq. (11). Ψ was obtained as follows:

$$\Psi = 0.081 \times 1 = 0.081.$$

Since Ψ is less than 0.25, the cooperative project was judged to be inviable.

Case 2

Since designers α and β can acquire new knowledge of items to which each of them gives the highest importance levels, the cooperative project was judged to be viable.

6.5.3. *Discussions of synergy effects of knowledge sharing*

The design solutions obtained by the cooperative project and the solutions obtained by each designer in isolation are now compared.

First, design solutions obtained by each designer before knowledge sharing are shown:

Design solutions of designer $\boldsymbol{\alpha}$

Materials that designer α can employ are mild steel and aluminum alloy, and he can use Motors No. 2 and No. 3. The constraints concerning discrete variables in the design optimization are:

$$(\tau, \theta_{\max}) \in \{(51, 3.14), (65, 1.99)\}$$
$$(\mathrm{E}, \rho) \in \{(2.1 \times 10^{11}, 7.8 \times 10^3), (6.9 \times 10^{10}, 2.7 \times 10^3)\}.$$

The feasible optimum design solution is on the Pareto optimum solution line shown by the broken line located on the upper left side of Fig. 21.

Design solutions of designer $\boldsymbol{\beta}$

Materials that designer β can employ are mild steel and FRP (fiber reinforced plastics), and he can use Motors No. 1 and No. 2. The constraints concerning discrete variables in the design optimization are:

$$(\tau, \theta_{\max}) \in \{(51, 3.14), (35, 1.99)\}$$
$$(\mathrm{E}, \rho) \in \{(2.1 \times 10^{11}, 7.8 \times 10^3), (6.9 \times 10^{10}, 2.7 \times 10^3), (5.9 \times 10^{10}, 2.0 \times 10^3)\}.$$

The feasible optimum design solution is on the Pareto optimum solution line shown by the broken line located on the right side of Fig. 21.

Next, design solutions obtained through knowledge sharing in the cooperative work of designers α and β are described. After knowledge sharing among designers α and β, the feasible materials are mild steel, aluminum alloy, and FRP, and the feasible motors are No. 1, 2, and 3. The constraints concerning discrete variables in

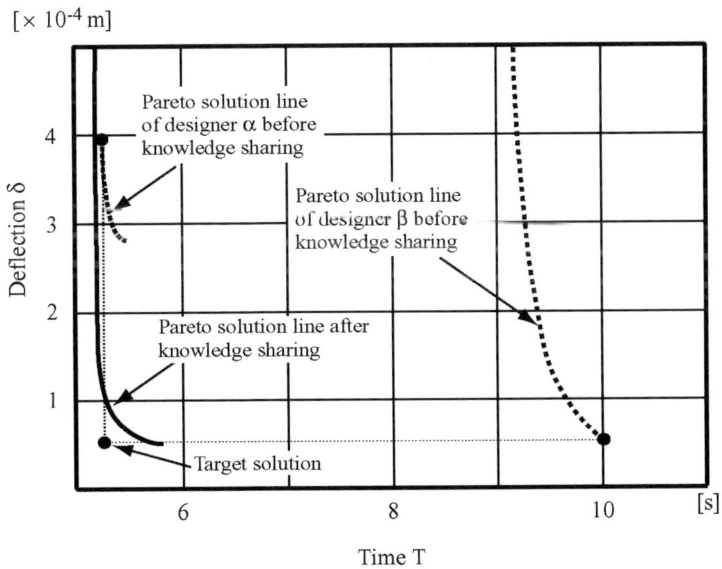

Fig. 21. Solution comparison of robot designs with two designers before and after knowledge sharing.

Table 5. Materials, and cross-sectional shapes of arms, and motors before and after knowledge sharing.

		Material	Cross-Sectional Shape	Motor
Before Knowledge	α	Aluminum alloy	Hollow	No. 3
Sharing	β	FRP	Solid	No. 1
After knowledge sharing		FRP	Hollow	No. 3

the design optimization are:

$$(\tau, \theta_{\max}) \in \{(51, 3.14), (35, 1.99), (65, 1.99)\}$$

$$(\mathrm{E}, \rho) \in \{(2.1 \times 10^{11}, 7.8 \times 10^{3}), (6.9 \times 10^{10}, 2.7 \times 10^{3}), (5.9 \times 10^{10}, 2.0 \times 10^{3})\}.$$

The Pareto optimum solution set obtained by solving the foregoing multiobjective optimization problem is shown on the solid line located on the lower left side of Fig. 21. The target point is an ideal point having two characteristic values each of which is the best feasible characteristic value realized by either designer α or β. The design solution on the Pareto optimum solution line after knowledge sharing is much closer to the target point than those of design solutions before knowledge sharing.

Table 5 shows the materials, the cross-sectional shapes of arms, and the motors that were used by designers α and β before knowledge sharing, as well as those used after knowledge sharing. As demonstrated in Table 5, a new design solution was obtained by knowledge sharing using designer α's knowledge of motors and arm shapes, and designer β's knowledge of materials. A superior product design having both high operational accuracy and high operational efficiency, which could not have been realized by an isolated designer, was thus obtained due to knowledge sharing.

7. Perspectives and Concluding Remarks

Presently, the ongoing development of industrial information systems is accelerating changes in industrial organizational structures as they evolve from pyramidal or hierarchical groupings to flat or networked types. In such circumstances, opportunities for cooperative work among divisions and/or among enterprises have already increased, and will continue to do so in the future. In order to realize the great benefits that these new industrial structures can offer, concurrent optimization techniques based on the collaboration concepts that were explained in this chapter will become increasingly important.

Due to increased communication among decision makers, such as that taking place during concurrent engineering product development, the development of new and valuable products and methods can be expected. A wide variety of people having both differing areas of knowledge and value structures can cooperatively evaluate product parameters from wider viewpoints. In addition, factors that had not been concurrently or simultaneously considered before can be included in the task of creating optimum designs.

Knowledge sharing among designers, groups or enterprises consisting of individuals having overlapping as well as unique knowledge based on their experience is a useful and beneficial strategy for realizing advanced product design solutions. In this chapter, although knowledge sharing among only two designers was discussed, the term "designers" can be replaced by "groups", "divisions", or "enterprises". These methodologies can be scaled up and efficiently applied to new cooperative projects between enterprises, and they can even be expanded to problems engendered by virtual enterprises, where different enterprises are linked together to accomplish a specific project, using networked systems.

References

1. A. Kusiak (ed.), *Concurrent Engineering — Automation, Tools, and Techniques* (John Wiley & Sons, New York, 1993).
2. C. T. Leondes (ed.), *Concurrent Engineering Techniques and Applications, Control and Dynamic Systems* **62** (Academic Press, San Diego, 1994).
3. H. R. Parsaei and W. G. Sullivan (eds.), *Concurrent Engineering — Contemporary Issues and Modern Design Tools* (Chapman & Hall, London, 1993).
4. G. Q. Huang, (ed.), *Design for X-Concurrent Engineering Imperatives* (Chapman & Hall, 1996).
5. W. Stadler (ed.), *Multicriteria Optimization in Engineering and in the Sciences* (Plenum Press, New York, 1988).
6. H. Eschenauer, J. Koski and A. Osyczka (eds.), *Multicriteria Design Optimization* (Springer-Verlag, Berlin, 1990).
7. M. Yoshimura, Concurrent optimization of product design and manufacture, *Concurrent Engineering*, eds., H. R. Parsaei and W. G. Sullivan (Chapman & Hall, London, 1993) 159–183.
8. J. L. Cohon, *Multiobjective Programming and Planning* (Academic Press, New York, 1978).
9. M. Yoshimura and A. Takeuchi, Multiphase decision-making method of integrated computer-aided design and manufacturing for machine products, *Int. J. Production Research* **31**, 11 (1991) 2603–2621.
10. M. Yoshimura and H. Kondo, Product design based on concurrent processing of design and manufacturing information by utility analysis, *Concurrent Engineering: Research and Applications* **4**, 4 (1996) 379–388.
11. M. Yoshimura and A. Takeuchi, Concurrent optimization of product design and manufacturing based on information of users' needs, *Concurrent Engineering: Research and Applications* **2**, 2 (1994) 33–44.
12. R. L. Keeney and H. Raiffa, *Decisions with Multiple Objectives: Preferences and Value Tradeoffs* (Wiley, New York, 1976).
13. M. Yoshimura and K. Yoshikawa, Synergy effects of sharing knowledge during cooperative product design, *Concurrent Engineering: Research and Applications* **6**, 1 (1998) 7–14.
14. T. L. Saaty, *The Analytic Hierarchy Process* (McGraw-Hill, Inc., 1989) 1–34.

CHAPTER 4

COMPUTER TECHNIQUES AND APPLICATIONS OF AUTOMATED PROCESS PLANNING IN MANUFACTURING SYSTEMS

KHALID A. ALDAKHILALLAH

Department of Management Information Systems and Production Management,
College of Business and Economics,
King Saud University,
PO Box 6033,
Al-Molaida Qassim, Saudi Arabia
E-mail: kasdalda@yahoo.com

R. RAMESH

Department of Management Science and Systems,
School of Management,
State University of New York at Buffalo,
Buffalo, NY 14260, USA

Manufacturing systems have become more and more sophisticated due to global competition. Hence, manufacturers have to satisfy conflicting demands for more product diversification, better product quality, improved productivity, and decreased cost. This trend in manufacturing systems and global competition have forced companies to adopt new sophisticated technologies by incorporating computer-based systems into their manufacturing systems. Evidence of this trend can be seen from the extent to which Computer-Aided Design (CAD), Computer-Aided Manufacturing (CAM), Computer-Integrated Manufacturing (CIM), Material Requirement Planning (MRP), Numerically Controlled (NC) Machines, Group Technology (GT), and Computer-Aided Process Planning (CAPP) are being used in industry. These computer techniques represent some of the developed systems that are used in today's world-class manufacturing systems. Therefore, the computer has been recognized by manufacturing firms as an important competitive weapon and a mean of survival. Current interest in manufacturing systems focuses heavily on integrating isolated computer-based systems into a unified system that handles and transforms information among these systems to facilitate a smooth production environment. A recent trend in integrating manufacturing system into a unified system is Computer-Aided Process Planning (CAPP). CAPP system bridges the gap between CAD system and CAM system. Therefore, CAPP system is a critical element in total system integration of automated design and manufacturing environment. This paper provides an overview of CAPP system and its approaches. Also, it presents the techniques that are used to integrate CAPP systems with other computer-based systems. Finally, some of the well-known CAPP systems will be discussed.

Keywords: Process planning; CAPP; CAD; CAM; CIM; AI.

1. Introduction

Computers have revolutionized the manufacturing industry for the past five decades. Computer usage in manufacturing systems started in the early 1950s when Massachusetts Institute of Technology (MIT) demonstrated a numerically controlled (NC) machine commencing the new era in manufacturing systems. Today's manufacturing firms depend heavily on computer-based systems in managing manufacturing information systems. As a result, computers are considered to be absolutely necessary tools for survival and for ensuring competitive advantages in the marketplace. The use of computer-based systems has several benefits. First, the computer-based systems provide a manufacturing firm the flexibility and speed to response to the customers' requirements. Second, these systems provide detailed and accurate analysis of data to strengthen a world-class manufacturing firm's ability to compete in a global market.

In recent years, manufacturing systems are becoming more and more sophisticated due to global competition. Therefore, in order to achieve a competitive edge, companies must be able to satisfy conflicting demands for greater product diversification, better product quality, as well as higher productivity at the lowest cost. This trend in the manufacturing system and the presence of global competition have forced companies to adopt new sophisticated technologies by incorporating computer-based systems into their manufacturing systems. Evidence of this trend can be seen from the extent to which computer-aided design (CAD), computer aided manufacturing (CAM), computer integrated manufacturing (CIM), material requirement planning (MRP), numerically controlled (NC) machines, group technology (GT), and computer aided process planning (CAPP) are being used in the industry. These computer techniques represent some of the developed systems that are used in today's world-class manufacturing systems. Therefore, the computer has been recognized by manufacturing firms as an important competitive weapon and a mean of survival.

Current interests in manufacturing systems focus heavily on integrating isolated computer-based systems into a unified system that handles and transforms information among these systems to facilitate a smooth production environment. This integration philosophy is called computer integrated manufacturing (CIM). CIM is defined as a closed-loop feedback system in which the functions of design and manufacturing are rationalized and coordinated using computers, networking, and information technology. CIM compromises a combination of several computer-based systems which represent the major elements of manufacturing system. The objective of CIM is to integrate isolated computer-based systems into the factory of the future, to deal effectively and efficiently with real-time analysis, planning, and control of the manufacturing process. There are several benefits of implementing CIM in a manufacturing environment. First, CIM improves the product quality, the firm's competitiveness, and its flexibility and responsiveness to market changes. Second, it reduces cycle time, flow time, lead time, and production cost. Third, it facilitates immediate access to up-to-date information from a central database. As a result, enhancements to the overall productivity and efficiency of a manufacturing system are possible from the use of the CIM system.

A recent trend in integrating manufacturing system into a unified system is computer aided process planning (CAPP) system. The CAPP system bridges the gap between CAD system and CAM system. Therefore, the CAPP system is a critical element in the total system integration of automated design and manufacturing environments.

Computer aided process planning (CAPP) integrates the automation of product design with that of manufacturing by linking the design representation of CAD systems with the manufacturing process representation of CAM systems. This integration has several benefits. First, the automation of process planning directly following an automated design stage results in consistent and accurate production plans. Second, integration reduces the workload on production planners and consequently decreases the planning cost and time. Third, it provides faster responses to changes in product design and/or in shopfloor status. Fourth, CAPP systems enable the firms to transfer a new product from concept into manufacturing in a short time. As a result, enhancements to the overall productivity of a manufacturing system are possible from the use of CAPP systems.

Two approaches to the design of CAPP systems are the variant and generative frameworks. The earliest work in CAPP was the variant approach which uses the group technology (GT) coding system to classify similar components into part families and generate standard process plans. On the other hand, the generative process planning approach develops plans automatically for a new product by synthesizing the process information for that product. This approach produces process plans by using different forms of decision logic, such as decision trees, for instance.

The automation of the development of manufacturing process plans starting from a product design can be seen as a two-stage process. The first stage deals with the determination of feasible production plans for the product by identifying the processing requirements from the design of the product. The second stage deals with the determination of an optimal production plan for the product by addressing issues such as shopfloor production planning, scheduling, and process control.

The first stage of automated process planning consists of three major tasks:

(i) recognition of the features of a product from a description of its design provided by a CAD system;
(ii) determination of the set of features that should be manufactured explicitly in any production plan for the product; and

(iii) determination of the set of feasible production plans, where each plan consists of a set of features (including those identified in (ii)) that when produced in some sequence would yield the target product.

The set of production plans is the input to the second stage of the automation process where each plan is organized into a production schedule using the shopfloor characteristics. Then, an optimal plan and its schedule are selected to achieve a desired production performance objective.

2. Approaches of Process Planning

Process planning is a function within a manufacturing system that involves translating the intention of design engineers to that of manufacturing engineers to produce a final product. In other words, process planning translates part design information from a description of its design provided by a design engineer into detailed work instructions to transform a part from its initial stage to its final stage. A detailed work instruction in process planning includes the following items: selection of appropriate machines, tools, machining processes, cutting tools, jigs and fixtures, determination of operation sequences, the generation of NC part programs, etc. Hence, process planning involves selection, calculation, and documentation and it represents a critical bridge between design and manufacturing. This task could be very complex and time consuming which requires a great deal of data.

In a traditional manufacturing environment, a process plan is generated by a process planner, who examines a part drawing to develop an efficient and feasible process plan and instructions to produce a part economically. This manual approach of process planning depends heavily on the knowledge and experience of the process planner to develop accurate, feasible, and consistent process plans. Therefore, the process planner must be able to manage and retrieve a great deal of data and documents to identify a process plan for a similar part and make the necessary modifications to the plan to produce the new part.

Process planning is an information-handling task which requires a significant amount of time and experience. Therefore, companies and researchers have attempted to automate the process planning task by using computer aided systems to handle the information required to generate a process plan. The automation of process planning represents a great challenge to companies and researchers since it requires modeling human intelligence on computers. As a result of intensive research, two approaches of the design of CAPP systems have emerged. These approaches are variant and generative frameworks.

2.1. *Manual approach*

Traditionally, process plans are generated manually by an experienced process planner who examines a part drawing to develop accurate and feasible process plans. This approach depends heavily on the knowledge and experience of the process planner. A process plan for a new part is created by two means. First, a process plan is created by recalling and identifying process plans for similar parts and then the necessary modifications are made to suit the new part. In this case, a process planner may use workbooks which are used to store and document information on previous plans or he may use his memory to retrieve similar process plans. Second, a process planner generates a unique process plan for the new part. This process is required since the necessary information is not documented or the requirements for the new part are not common to the existing family parts. This approach is considered to be suitable for small firms with very limited number of process plans to prepare. However, as the number of process plans increases, the need for a computerized system to perform process planning task increases.

The manual approach has several disadvantages. First, process plans that are generated for the same part by different planners will be usually different. This reflects the fact that most of these plans, if not all, are not efficient manufacturing methods. Second, a process planner may develop a process plan for a part during a current manufacturing program which might be quite different from a plan that was developed for the same part in a previous manufacturing program. This results in huge wastage of time and effort and produces inconsistent process plans. Third, personal experience and preferences of the process planner are reflected on the process plans generated. Finally, this approach is labor-intensive, time-consuming, and very costly in the long run.

2.2. *Variant approach*

The early attempt to automate process planning function in a manufacturing environment was the development of the variant CAPP system. This approach represents an extension of the manual approach of process planning in which it requires recalling, identifying, and retrieving previous plans for similar parts. Then, the retrieved plan usually needs to be modified for the new part. The computer assists the process planner by providing an efficient and highly fast system for data management, retrieval, and the editing of process plans. Therefore, the variant approach of CAPP system is a computerized database retrieval approach.

The variant approach uses group technology (GT) coding system to classify similar components into families and generates standard process plans. Hence, the development of the variant approach requires coding and classifying existing parts into part families. Then, a standard process plan is prepared for each part family. The part family information and the corresponding standard process plans are stored in a database.

Figure 1 shows the variant approach's development. To retrieve a standard process plan suitable for a new part, the new part has to be coded and classified into a part family, The process planner has to develop a new plan for the new part if this part cannot be classified into a family. Hence, the variant approach utilizes the fast storage and retrieval capabilities of computers and provides an interactive

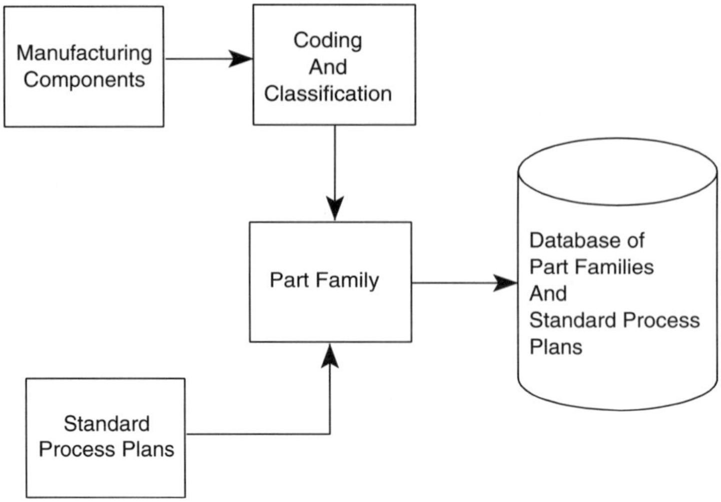

Fig. 1. The variant approach development.

environment between the planner and the computer. Figure 2 shows the operating sequences for creating a process plan for a part using the variant approach.

There are several advantages associated with the variant approach. First, the classification and coding of parts into family facilitate the standardization of process plans which results in significantly greater consistency in process plans. Second, using speed and accuracy of computers increase information management capabilities. Third, this approach significantly reduces time and labor required to create process plans. However, the variant approach has the following disadvantages. First, it requires an experienced process planner to maintain the system. Second, the efficiency and feasibility of process plans depend on the knowledge and experience of the process planner.

2.3. *Generative approach*

The generative process planning approach develops plans automatically for a new product by synthesizing the process information for that product. This approach produces process plans by uniquely determining processing decisions to convert a product from its initial state to its final state using different forms of decision logic, technology algorithm, and geometry-based data. In this approach, a process plan is developed using manufacturing rules and information on equipments and tools that are available in a manufacturing database system. Hence, a unique process plan is generated for each product without human intervention. In order to generate a process plan, it is necessary to analyze the geometrical information of the product under consideration.

The analysis of the product geometrical information is the input to the generative process planning system. This analysis can be accomplished by either a text

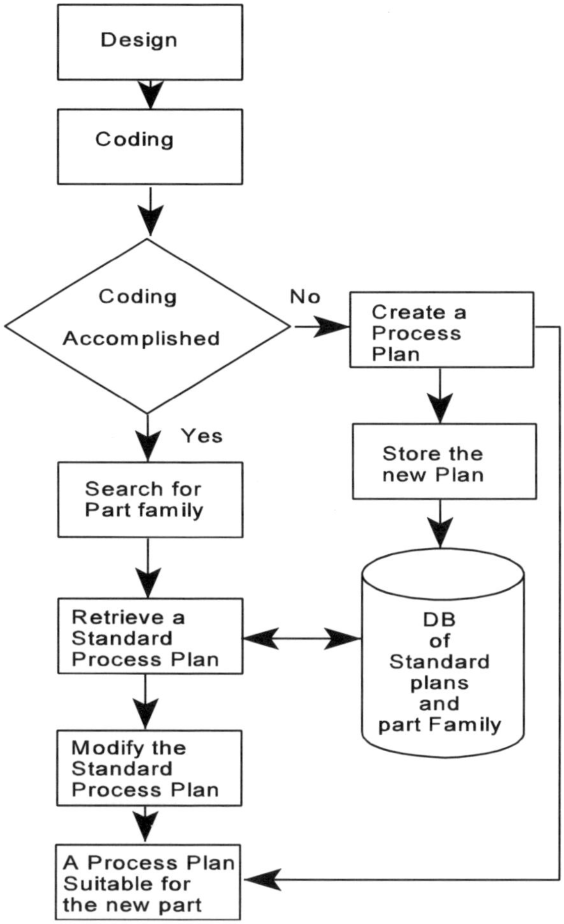

Fig. 2. Operating sequence of a variant CAPP system.

input through the user or a graphic input from the CAD data. In the text input, the process planner answers a number of questions concerning the part characteristics to translate the part information into computer interpretable data. On the other hand, the part data are gathered directly from the CAD data using feature recognition or feature-based design techniques to translate the CAD product representations into information as required by CAPP. The objective of the part analysis is to extract manufacturing features to simplify process planning function. Several approaches have been developed for feature extraction from 2D or 3D CAD databases.

The analysis of product information is the input to the second stage of the generative process planning approach where a set of programs transforms this information into detailed process plan instruction. The set of programs consists of decision logic (e.g. decision table, decision tree), formula, and technological algorithms. The aim of the program is to compare product information and requirements to manufacturing capabilities and shopfloor capacity and then generate a process plan.

The generative process planning approach has the following advantages. First, the generative approach produces process plans automatically for a new part without human intervention. Second, it produces consistent process plans as soon as the information of the new part is available. Third, time and labor that are required to generate a process plan are significantly reduced. However, this approach has the following shortcomings. First, the generative process planning approach requires a large database, which contains data on available tools, jigs and fixtures, machines, etc. Second, it requires very complex algorithms and decision logic to capture the manufacturing knowledge. Third, the system is normally developed for a specific manufacturing environment. Finally, the development of a generative process planning system requires a thorough understanding of manufacturing knowledge.

3. Artificial Intelligence in CAPP Systems

Artificial Intelligence (AI) is a computer science field that is aimed to create a device that performs specific functions (reasoning, planning, problem solving, etc.) automatically and intelligently. These functions require normal human intelligence to be performed. AI has been applied in the exploration of the nature of human intelligence or in the determination of how computers can be used to solve specific problems. As a result of intensive research in this field, many concepts and techniques have been developed by AI researchers. Some of these concepts and techniques are natural language, robotics, exploratory programming, improved human interfaces, expert systems, and scheduling. AI-based CAPP systems have been considered as a branch of expert systems.

An expert system is a program that consists of concepts, procedures, and techniques. These techniques are used to design and develop system that uses knowledge and inference techniques to analyze and solve problems. The knowledge of human experts is usually represented by the following tools: inductive tools, simple rule-based tools, structured rule-based tools, hybrid tools, and domain tools. The most common approach of knowledge representation in the CAPP systems is rule-based tools. To develop an effective expert system, the knowledge representation must be expressive, unambiguous, effective, clear, and correct. There are two types of knowledge involved in automated process planning systems: declarative knowledge and procedural knowledge.[1]

The construction of a knowledge base using declarative knowledge approach is accomplished by adding sentences one by one which represent the designer's knowledge and understanding of the environment. These sentences are expressed in a language called a knowledge representation language, which is defined by two aspects: syntax and semantics. On the other hand, the procedural knowledge can be represented by If–Then statements. These statements are call production rules. A system consisting of a set of production rules is called a production system.

The production system uses implications as its primary representation and the consequences of these implications are interpreted as action recommendations.

The automation of process planning function in a manufacturing system using AI-based techniques has played a dominant role in recent years. Since process planning is a function in a manufacturing system that requires a great deal of human intelligence, knowledge, and expertise, AI has a great potential in automating the process planning function using knowledge-based expert systems. Process planning is a decision-making mechanism that consists of procedures to transform a part design into a final product. On the other hand, planning in AI is a problem-solving task that determines a sequence of actions leading to a desirable state. Process planning consists of the following elements: initial state (part description), goal state (final product), and a set of operators to transform the initial state into the goal state. Several knowledge-based systems have been developed to automate process planning. Some of the well-known systems are GARI,[2] TOM,[3] HI-MAPP,[4] and CIPPS.[5,6]

AI-based CAPP systems are designed to capture, represent, organize, and utilize pervious knowledge in the manufacturing domain. Then, this knowledge is used to create a formal representation of objects and relations in this manufacturing domain. The process of acquiring knowledge for a particular domain is called knowledge engineering. A knowledge engineer must understand the manufacturing domain to be able to represent the important objects and relationships in this domain effectively and efficiently. In AI-based techniques, human expert are normally the source of knowledge required in a particular domain. However, due to the complexity of the manufacturing processes, knowledge in manufacturing should be obtained through not only human experts but also through manufacturing data and shopfloor data. Therefore, in order to successfully develop and implement an AI-based generative process planning system, one needs to identify, fully understand, and capture the manufacturing logic used by the process planners and the manufacturing and shopfloor data. In this case, AI-based techniques will be used to their full potential in developing CAPP systems that provides integration capabilities within a manufacturing environment in the future.

4. Integration with CIM

One of the most significant trends in the manufacturing industry in recent years is the attempt to integrate manufacturing functions into a unified system. This integration provides companies competitive advantages in the basis of product diversification, better quality, improved productivity, and decreased cost and time. This integrative system is call computer integrated manufacturing (CIM). The CIM system is a closed-loop feedback system in which many complexes and interrelated manufacturing functions are rationalized and coordinated using computers, networking, and information technology. The fundamental goal of the CIM system is to unify the diverse areas of design, engineering, production processes, inventory, sales and purchasing, and accounting into a single interactive closed-loop control system. Hence, CIM helps a manufacturing firm to perform more flexibly and efficiently in today's global market through the integration of all business functions.

This trend in manufacturing systems has been facilitated by the rapid development in computing technologies and the appreciation of the effects of computerization on manufacturing performance. CIM has demonstrated great potential for improving manufacturing capabilities, effectiveness, and efficiencies. The technological components of CIM consist of several essential key elements such as GT, CAD, CAM, CAPP, etc. Hence, CIM can be seen as an umbrella that covers these technological components.

CAPP system is a critical element in total system integration, hence, it has a key role to play in a CIM system. CAPP system emerges as a key factor in CAD/CAM integration by linking the design representation of the CAD system with the manufacturing representation of the CAM system. There are many approaches of CAPP systems that have been developed to integrate design and manufacturing. Therefore, CAPP system is an intermediate system that translates the intention of design engineers to that of manufacturing engineers to produce a final product.

4.1. *CAD/CAPP integration*

The objective of CAD system is to create, analyze, and optimize the product design. Hence, CAD part representations provide an extensive part description data in terms of geometry, tolerance information, material type, and information necessary to analyze the part. The CAD part representation is stored in a CAD database as a 2D or 3D geometric representation of the part. There are several part modeling techniques: wireframe, surface modelers, constructive solid geometry (CSG), boundary representation (BRep), and spatial occupancy enumeration (SOE).

After the part has been designed, a solid modeler allows the designer to store the part data in a database containing the geometry and topology of the part. CSG and BRep represent the two major forms of storage. CSG stores the part data as a tree where the leaves are primitives and the nodes are Boolean operators. In BRep, linked lists of all vertices, edges, and faces which incorporate the geometry and topology of the part are maintained.[7] The part data that is stored in a CAD database represents an essential component for performing process planning, manufacturing planning, and downstream manufacturing functions. However, the CAD part representation in a part modeler differs from the type of information required by the CAPP systems. This information is implicitly embedded within the part representation data and needs to be automatically extracted and organized in a form suitable for use in a CAPP system.

Therefore, the automation of process planning requires the development of a method to infer about the part and reasoning geometrically to extract information from the CAD database automatically. Several approaches have been developed for feature recognition and extraction from 2D or 3D CAD databases. Comprehensive reviews of these approaches are cited in Refs. 8–10. Joshi and Chang[11] have developed an attributed adjacency graph (AAG) to recognize features from boundary representation (BRep) of CAD data. In the approach of Vandenbrande,[8] the product information is processed by production rules to generate hints for feature presence using a generate-and-test strategy. The promising hints are processed to generate the largest possible feature volume which does not intrude into the product and is consistent with the available data. Kao[10] has developed a super relation graph (SRG) system for feature recognition that employs artificial neural networks. The objective of this system is to recognize and extract prismatic features from 3D CAD databases. The SRG has been implemented by Gallagher.[12] This system results in the decomposition of the cavity volume into a set of volumetric primitive features.

In addition, feature-based design is currently being used to facilitate the integration of CAD and CAPP systems. The feature-based design allows the designer to fully utilize features throughout the product life cycle. Hence, feature-based design has the potential of supporting the design process better than current CAD systems do. There are several advantages of a feature-based design. First, it can speed up the design process and provide means of standardization. Second, it reduces the cost of designing process and designing time. Third, it improves the link between the CAD and CAPP systems. A comprehensive review of feature-based design is given in Ref. 13.

The use of feature recognition techniques or feature-based designs to transfer model data between various systems is still quite problematic. As a result, there have been efforts to develop design representation standards as the basis for complete product definition. The data transfer standards are product definition data interface (PDDI), initial graphics exchange standard (IGES), product data exchange specification (PDES), and standard for exchange product model data (STEP). In summary, in order to fully integrate the CAD and CAPP systems, there must be a mechanism to transform the CAD part description data in a form suitable for use in CAPP system in terms of manufacturable features. The manufacturable features represent the key to generate a process plan for a product. Therefore, to implement a CIM system, the CAD and CAPP systems have to be integrated.

4.2. *CAPP/CAM integration*

Computer aided manufacturing (CAM) has been defined by CAM-I as "the effective utilization of computer technology in management, control, and operations of the manufacturing through either direct or indirect computer interface with physical and human resources of the company".[14] Computer technologies have revolutionized manufacturing system since the demonstration of numerically controlled (NC) machines in 1950s. In today's manufacturing environment, computers are being used in several functions in manufacturing such as manufacturing engineering, material handling, scheduling, inventory, MRP, and etc. CAM system can be seen as an umbrella that covers all the computerized manufacturing functions through a computer network. Therefore, the architecture of the manufacturing system can be described as a series of activities that are connected through a network.

As mentioned earlier, process planning is a function within a manufacturing system that translates part design information from a description of its design provided by a design engineer into a detailed work instruction to transform a part from its initial stage to its final stage. Hence, process planning bridges the gap between the CAD and the CAM system. However, most CAPP systems are developed without consideration to downstream manufacturing functions status information. As a result, process plans are developed in isolation of shopfloor status information such as random disturbances or planned production changes. Consequently, a large number of process plans have to be altered to cope with these disturbances.

There have been attempts to integrate process planning and production scheduling. This integration is essential to achieve eventually a totally integrated manufacturing system. Several approaches for integrating process planning and production scheduling have been developed. These approaches are nonlinear process planning, flexible process planning, closed loop process planning, dynamic process planning, alternative process planning, and just-in-time process planning.[15] The nonlinear process planning approach (also referred to as an alternative process planning) creates all possible plans and then prioritizes them based on manufacturing criteria. The scheduler examines these plans based on their priority until a suitable plan is achieved. In the approach of closed loop process planning (also referred to as a dynamic process planning), plans are generated by means of a dynamic feedback from production scheduling. In the just-in-time process planning (also, referred to as a distributed process planning), process planning and production scheduling are performed simultaneously.

A careful examination of the above-mentioned approaches of integrating process planning and production scheduling reveals that the only truly integrated approach is the just-in-time approach. However, the rest of them provide interfacing between process planning and production scheduling. The difference between interfacing and integration as stated by Ham and Lu[16] is that interfacing is achieved at the result level while integration is addressed at the task level.

5. A Survey of CAPP Systems

In the following discussion, we survey some of the important CAPP systems proposed and developed at both the industry and research institutions.

5.1. *CAPP*

The need for database information control of the manufacturing system was recognized by Computer Aided Manufacturing-International Inc. (CAM-I) organizers in 1972.[17] As a result, CAM-I sponsored the process planning program as one of its five project program. The goal of the process planning program is to automate manufacturing process planning and to integrate CAD and CAM with geometric modeling capabilities. McDonnell Douglass Automation Company was chosen to be the project's development contractor. McAuto which is a division of McDonnell Douglass Automation Company designed and implemented the CAPP system. This automated system was demonstrated in St. Louis, Missouri in 1976. It was called the machined part process planning creator module and it is directed towards machined parts.

The objectives in creating this system are: (1) to examine the feasibility of computer systems for process planning and (2) to determine a planning system's benefits and shortcomings. The CAM-I CAPP variant system is a database management system which was written in ANSI standard FORTRAN. The system logic is based on GT methods of classifying and coding parts in a structured database. A new part is coded with as many as 36 alphanumeric code and then compared with the existing code system to extract a standard process plan for the new part. This plan can be edited partially or totally or accepted as it is.

5.2. *GENPLAN*

The GENPLAN (GENerative process PLANning) system was developed by Lockheed Georgia.[18] This system contains a manufacturing-technology database that was developed through an extensive analysis of previous process plans. This database is accessed through a group technology (GT) coding system. The coding system covers part geometry, size, and manufacturing processes. GENPLAN consists of the following procedures: (a) the determination of the sequence of operations, (b) the selection of machine tools, (c) the calculation of machining time.

5.3. *GARI*

GARI is an AI-based process planning system developed at the University of Grenoble in France.[2] The system uses a set of production rules as a representation of its knowledge base. In GARI, a part is represented to the process planning module in terms of a set of form features (e.g. holes, notches, etc.) which includes geometrical and technological information. The system provides the capability of backtracking mechanism from any of the intermediate stages of the process planning development to provide the necessary revisions. It assigns weights to different pieces of advices at each stage of the process planning development to resolve conflicts that appear. The system is written in MACLISP and operates on CII-Honeywell Bull HB-68 computer under the MULTICS operating system.

5.4. *CMPP*

CMPP (Computer Managed Process Planning) is a generative process planning system that was developed by the United Technologies Research Center in cooperation with the US Army Missile Command.[19] The system makes process decisions automatically with interactive capability that allows the planner to monitor the development of the process plans. CMPP was developed for machined cylindrical parts that are characterized by tight tolerance and complex manufacturing processes. The system performs a set of processes for cylindrical features (e.g. turning) and non-cylindrical features (e.g. milling). The three functional areas of CMPP are as follows: the building and maintenance of a manufacturing database, the definition of part model, and the generation of process planning. The manufacturing database consists of manufacturing logic and manufacturing resources. Manufacturing logic is defined by a process decision model which is written in the computer process planning language (COPPL). The manufacturing resource contains information on available machines and tools.

In this system, the part design description is entered into the system through the part model definition. After the part description has been entered, the system performs four functions. First, the operation sequences in a summary format is generated. Second, the dimension reference surfaces for each cut in each operation is selected. Third, the analysis and determination of machining dimensions, tolerance, and stock removals for each surface cut in each operation take place. Finally, the process plan documentation is generated.

Liao *et al.*[20] have modified the CMPP system to achieve CAPP/scheduling integration. They have modified the process decision model and the machine tool file by incorporating a scheduling criteria (mean flow time and number of jobs tardy) to perform machine selection.

5.5. *TOM*

TOM (Technostructure of machining) is a rule-based expert system developed at the University of Tokyo.[3] TOM translates a part design and data from the COMPAC CAD system using the EXAPT part program for preparing inputs for a CAPP system. Also, the system allows the user to enter the part design description directly. TOM uses production rules as its knowledge representation scheme about machining operations, sequencing, and geometry of a part. TOM employs a backtracking search mechanism to generate a process plan. The search is performed with Alpha-Beta (a well known AI technique) strategy to significantly increase the search efficiency. TOM starts by obtaining the part information from the EXAPT part program, then it executes the production rules using a backward chaining mechanism. As a result, a new intermediate geometry is obtained in which a few rules would be applicable. This system is written in PASCAL and runs on VAX and it handles holes exclusively.

5.6. *HI-MAPP*

HI-MAPP (Hierarchical and Intelligent Manufacturing Automated Process Planning) is an artificial intelligence (AI) based process planner developed at the

University of Tokyo.[4] HI-MAPP runs in InterLips under the VNS 4.1 operating system on a DEC VAX/750 computer. In this system, a part is represented using a set of features which represent geometric entities such as notches, trapped holes, and faces. Then, each feature is treated as a working element in the machining process. The knowledge base in HI-MAPP consists of 45 production rules that are classified into four categories. The first category consists of rules that define a selected process such as milling operation. The second category consists of rules that recommend the type of cut. The third category contains rules that recommend the machine, and the final category consists of rules that provide for miscellaneous action which can be defined by the user. HI-MAPP then applies the hierarchical and nonlinear planning concepts.

5.7. *KAPPS*

The KAPPS (Know-How and Knowledge Assisted Production Planning System) system incorporates the know-how of experienced production engineers into a CAPP system.[21] KAPPS consists of four subsystems as follows: (1) the CAD interface and user input; (2) the decision making subsystem; (3) the know-how and databases; (4) the know-how acquisition. The CAD interface and user input subsystem translates the part model data to the list type data structure. The data structure includes face numbers, geometric features, dimensions, related positions among faces, and surface roughness. The decision subsystem applies the related know-how and data to solve a set of problems by generating a search tree and forward reasoning method. The know-how base subsystem represents and stores the know-how and knowledge that are obtained through the know-how acquisition subsystem. The know-how acquisition subsystem receives and retains the know-how and knowledge of an experienced production engineer to the know-how base interactively. The decision making engine procedures consist of seven steps as follows: (a) reads data from CAD database; (b) makes and updates a temporary file; (c) defines a problem to be solved; (d) decides a sequence of procedures; (e) makes a feasible solution using a search tree; (f) evaluates the solution; and (g) prints the final solution. The frame method is written in COMMON-LISP language.

5.8. *Propel*

Propel is a feature-oriented generative process planning system for orthomorphic and ono-orthomorphic prismatic parts of average complexity.[22] The principle of Propel is based on two AI techniques: an opportunistic combination of plans planning strategy and a compromising algorithm for resolving contradiction when they arise. The opportunistic combination of plans consists of two steps. First, the problem (machining of a part) is broken down into subproblems (machining of the features), then the subsolution for each subproblem is obtained. Second, the subsolutions are combined to form a global solution (process plan for the part). The compromising algorithm is coupled with opportunistic combination of plans planning strategy to resolve contradictions when they appear. The input to the system consists of part description, initialization of knowledge base, and constraints in the knowledge base. The part description requires the establishment of a hierarchy of features and the hierarchy of relationships among these features. The production means description is defined by the available machines and tools that are represented by a hierarchy of machine types and hierarchy of tool types. The initialization of knowledge base consists of rules that represent the manufacturing processes. This system is written in COMMON-LIPS language and runs on a SUN 3/160 workstation under UNIX.

5.9. *Turbo-CAPP*

Turbo-CAPP[23] is an intelligent process planning system in which a problem-solving knowledge is represented as frames or production rules and is stored in three planning layers: layer of facts, layer of inference rules, and layer of meta-knowledge. This system consists of five modules as follows: (1) machine surface identification; (2) process selection and sequence; (3) NC code generation; (4) knowledge acquisition; and (5) database management. The function of the machine surface identification module is to extract part information from a 2D design system. The process selection and sequence module has two submodules as follows: the knowledge base submodule and the inference engine submodule. The NC code generation module develops an NC program for each part based on its geometric features. The knowledge acquisition module consists of three routines that perform the knowledge acquisition. These routines are tolerance input, machine description, and process manipulation. The database management module controls the relationships among the system's modules.

The system starts with extracting geometric entities in terms of surface features from the part description provided by the 2D CAD system. Then, the features and the qualification data are input into the process selection; the sequence module and the NC code generation module to develop alternative process plans and corresponding NC codes. This system is implemented on IBM PCs in PROLOG.

5.10. *IPPM*

IPPM (Integrated Process Planning Model) integrates the process planning function and the scheduling function.[15] The system is based on the distributed process planning concept. IPPM consists of three modules as follows: process planning module, production scheduling module, and decision making module. The system uses a real time feedback mechanism to integrate these modules. The three modules are integrated in three levels as follows: preplanning level, decision making level, and functional integration level. At the preplanning level, the process planning module performs the feature reasoning, the process recognition, and the setups determination procedures. The feature reasoning extracts the part's features by analyzing the part design description. The process recognition procedure performs the machine selection and equipment information from the shopfloor to provide availability and flexibility for machines selection. The setups determination selects the adequate and optimal setups based on feature recognition and automated tolerance analysis. The production scheduling module provides the available equipment information at the preplanning level. At the decision making level, the process planning module consists of the following: (1) machine selection which is based on real time feedback information; (2) tooling and fixturing selection and (3) time calculation. At this level, scheduling constraints are involved rather than process planning rules. At the functional integration level, a detailed process planning and a detailed production scheduling are performed simultaneously. To achieve a truly integrated process planning and production scheduling, their constraints should be considered simultaneously.

6. CIPPS

The CIPPS (Computer-Integrated Process Planning and Scheduling) system represents the design architecture and the operational framework of a CAPP system that also incorporates production scheduling.[5] The objectives of the CIPPS system are as follows: (1) to recognize the features of a product from its description provided by the CAD system; (2) to identify the sets of features for which production must be explicitly planned; (3) to determine an efficient and feasible process plan to manufacture a product; (4) to generate an efficient and feasible cyclic production schedule from the process plan; (5) to provide the design and manufacturing engineers with the necessary feedback to appropriately and fully evaluate a design and ensure that the product can be manufactured in a cost-effective manner; (6) to equip each module of the system with intelligent capabilities to react to random shopfloor disturbances as well as planned production changes; and (7) to seamlessly integrate CIPPS with other automation processes and systems within the framework of a computer-integrated manufacturing (CIM) environment. The CIPPS system consists of four integrated modules as follows: the super relation graph (SRG)[10] for automated feature recognition; the cover set model (CSM) for the determination of minimal cover sets of product features;[24] the cover set planning and scheduling algorithm (CSPS) for the determination of an efficient and feasible process plan;[25] and the cover set cyclic scheduling algorithm (CSCS) for the generation of an efficient and feasible cyclic production schedule for a production schedule for a production plan.[6] Objectives (1) through (4) are individually achieved using the modules SRG, CSM, CSPS and CSCS, respectively. The overall framework of operation CIPPS system addresses objectives (5)–(7).

The SRG module is employed to recognize polyhedral depression features and extracts interacting prismatic features from 3D CAD databases. The CSM module follows from the SRG module and determines all the sets of features that have minimum cardinality and it covers all the features of a product recognized by the SRG module. These sets are known as 'minimal cover sets' (MCS). The CSM module consists of four algorithms operating in the following sequence: perfect feature creation (PFC); solid feature creation (SFC); cover sets determination (CSD); and minimal cover sets (MCS) determination. The CSPS module identifies a set of feasible process plans and extracts an efficient process plan from this set. This module consists of two submodules: process planner (PP) and process scheduler (PS). The submodule PP determines a set of feasible pairwise feature production plans to manufacture the product. The feasibility of a process plan is governed by a set of geometric and technological constraints on the production process. The submodule PS determines an efficient overall process plan from the feasible set identified by PP using a Hamiltonian path heuristic for the optimization problem underlying the feature sequencing process. The CSCS module determines an efficient and feasible cyclic production schedule for a job shop in which a product is produced on a set of machines using a given predetermined sequence of features specified in the process plan determined from the CSPS module.

The system framework specified three modes of CIPPS operations: the dynamic support for design decisions (DSDD); the runtime intelligent operational control (IOC); and the data consolidation and integration (DCI) mode. In the DSDD mode, CIPPS supports decision making in the design process. In the IOC mode, the automatic intelligent shopfloor management is facilitated when changes occur in the environment. In the DCI mode, CIPPS is interfaced and integrated with other functions in a manufacturing environment.

6.1. *Architecture of CIPPS system*

The framework shown in Fig. 3 depicts the modules, their interfaces, the transient files and the master files in the CIPPS system. It also shows the information feedback flow from the modules to the design and manufacturing engineers who have the responsibility of ensuring that a design can be manufactured in a cost-effective manner. This is facilitated by the feedback on the projected process plans and schedules at the design stage itself by the CIPPS system.

At the design stage, the design and manufacturing engineers should address the geometry, the material properties and the specifications for the product. The geometric data of the product is the input to the SRG module. The material and specification data are inputs to the CSPS module. After a product is designed, the CIPPS system develops a product database which contains the information provided by the CAD database. The CIPPS database is progressively built by its successive modules during the course of design and process planning.

6.2. *Operations of the CIPPS system*

The CIPPS system framework is designed to provide (1) dynamic support for design decisions (DSDD), (2) runtime intelligent operational control (IOC), and (3) data

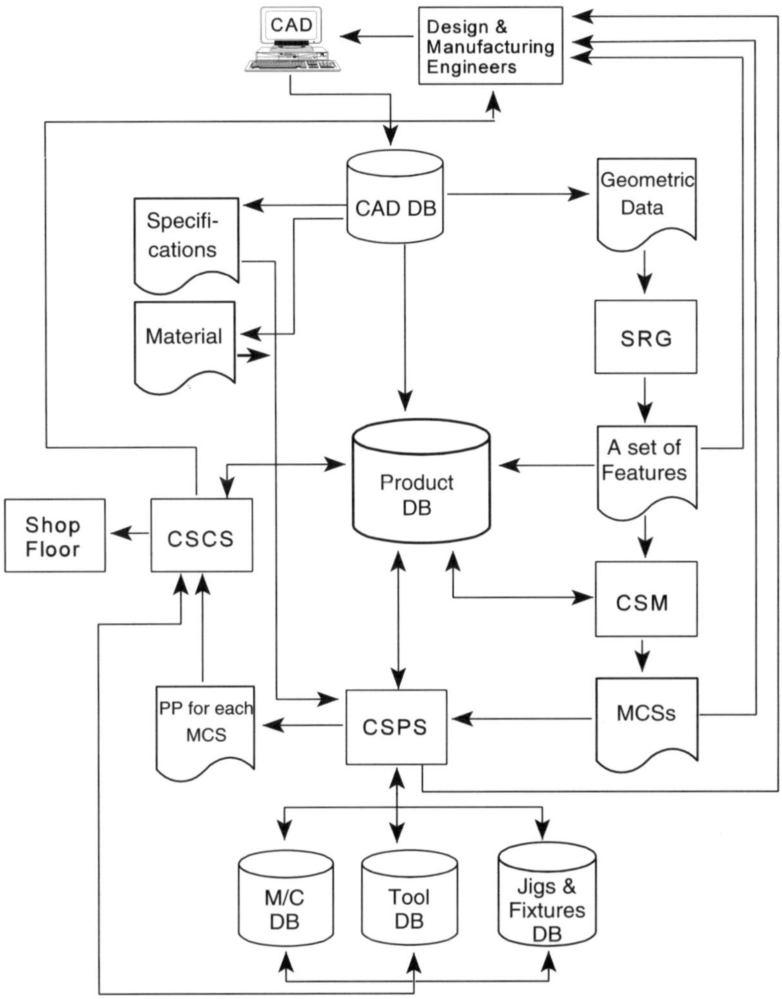

Fig. 3 The overall architecture of the CIPPS system.

consolidation and integration (DCI). The system is operated in three modes, corresponding to the support functions as stated above. We develop the structure of these modalities in the following discussion.

6.2.1. *Dynamic support for design decisions (DSDD)*

In the DSDD mode, CIPPS provides the design and manufacturing engineers with the necessary planning/scheduling feedback information in real time to facilitate the evaluation of design and process at various stages of product/process development. As the product is being designed, the design and manufacturing engineers work together to ensure that the product can be manufactured in a cost-effective

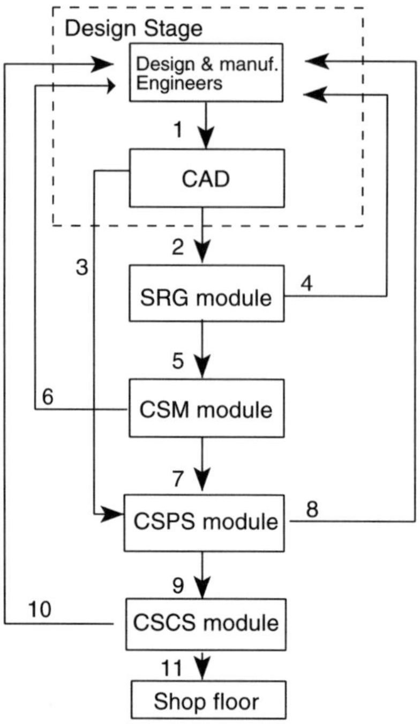

Keys:

1- The designer's concept of the product.

2- Geometric data of the product.

3- Material type and specifications of the designed product.

4- Feedback from the SRG module to the design and manufacturing engineers on the validity of the product's features.

5- A set of recognized features.

6- Feedback from the CSM to the design and manufacturing engineers on the MCSs of the designed product.

7- Minimum cover sets.

8- Feedback on the machine, tool, jigs and fixtures selected and the cost of production.

9- An efficient process plan and the associate costs and time for each MCS.

10- Feedback on the total cost of manufacturing an order for each MCS.

11- Production order to the shop floor.

Fig. 4. Dynamic support for design decisions.

manner. CIPPS operates in the DSDD mode in five stages: product design, feature recognition, MCS determination, the development of an efficient process plan for each MCS, and the generation of an efficient and feasible cyclic production schedule. Figure 4 shows these stages and the flow of information among them.

6.2.2. *Intelligent operational control (IOC)*

During the production process, random shopfloor disturbances or planned production changes may occur that will affect the cyclic production schedule. The CIPPS modules have intelligent capabilities to react to these disturbances through the intelligent operational control (IOC) mode of the system operation. This mode is triggered by the random or planned events at the shopfloor level. In this mode, the system backtracks from the cyclic production schedule to the earlier stages of planning and ultimately to the product design if necessary to handle the shopfloor disturbances. Figure 5 shows the flow of information among the modules of CIPPS in the IOC mode.

6.2.3. *Data consolidation and integration (DCI)*

The CIPPS system can be integrated with other functions of a manufacturing system into a totally automated manufacturing environment. CIPPS creates a product

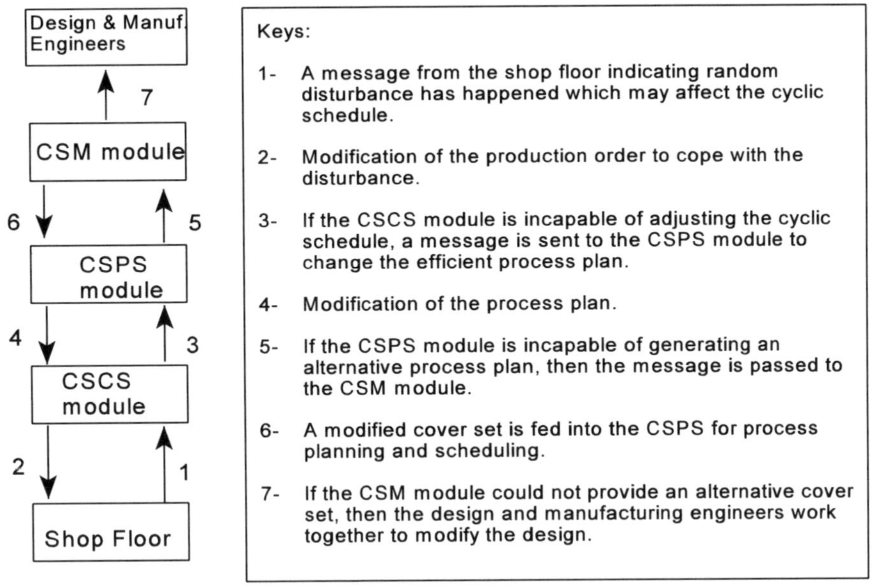

Fig. 5. Intelligent operational control.

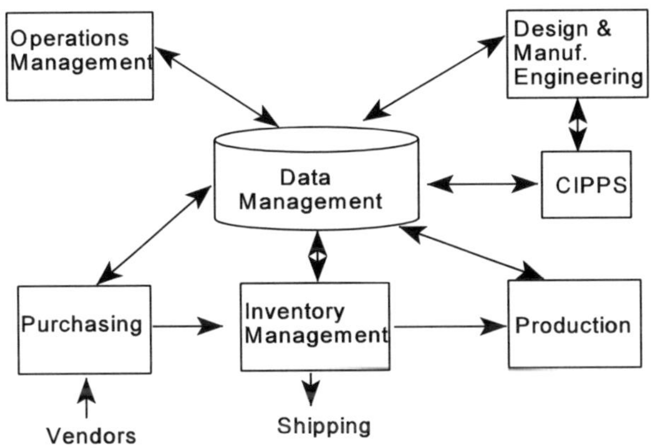

Fig. 6. CIPPS integrated with other systems.

database that can be consolidated and integrated into a computer integrated manufacturing (CIM) system. A typical manufacturing environment consists of the following departments: design engineering, manufacturing engineering, operations management, production, inventory management and purchasing. Figure 6 shows the structure of the integrated manufacturing environment. The CIPPS system is interfaced and integrated with the other functions of a manufacturing environment. These functions interact through a data management system.

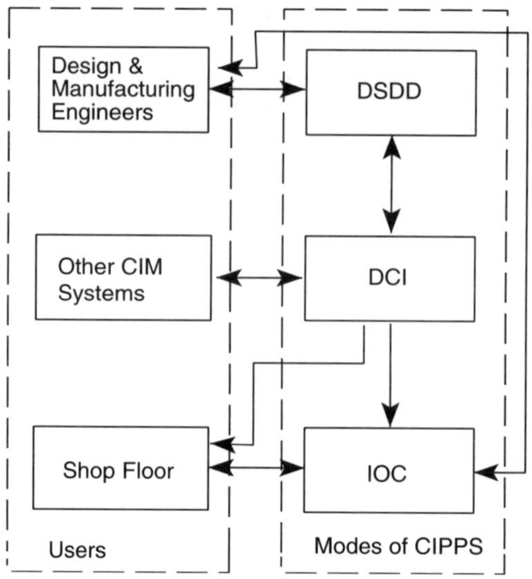

Fig. 7. Information flows in a CIPPS integrated CIM environment.

Figure 7 shows the flow of information among the departments of a manufacturing environment. The design and manufacturing engineers interact with CIPPS in its DSDD mode in an iterative manner until satisfactory design, features, cover sets, and process plans are obtained. During this process, the product database is updated to reflect the status of the designed product. After the product is designed and approved for production, the CIPPS system produces documents which consist of: a detailed process plan description, a work order to the shopfloor, material requisition, and the total cost of production. These documents are maintained in a database by CIPPS in its DCI mode. Other CIM components and departments in a manufacturing environment interact with CIPPS in the DCI mode for data storage and retrieval purposes. The production department retrieves the work orders and determines the appropriate time to start production. The inventory management department handles the material requisition and places an order to vendors. The operation management evaluates the total cost of production to determine profitability. The shopfloor personnel and system interact with CIPPS in its IOC mode during production to handle random disturbances or planned production changes. The IOC mode also provides feedback to the design and manufacturing engineers on any required changes.

References

1. T. C. Chang, and R. A. Wysk, *An Introduction to Automated Process Planning Systems* (Englewood Cliffs, NJ: Prentice-Hall, Inc., 1985).
2. Y. Descote and J. C. Latombe, GARI: A problem solver that plans how to machine parts, *Proc. 7th Int. Joint Conference On Artificial Intelligence*, Vancouver, Canada, August 1981.

3. K. Matsushima, N. Okada and T. Sata, The integration of CAD and CAM by application of artificial intelligence, *Annals of the CIRP* **31**, 1 (1982).
4. H. R. Berenji and B. Khoshnevis, Use of artificial intelligence in automated process planning, *Computer in Mechanical Engineering* (1986) 47–55.
5. K. A. Aldakhilallah, *An Integrated Framework for Automated Process Planning and Scheduling*, PhD Dissertation, State University of New York at Buffalo, USA, 1997.
6. K. A. Aldakhilallah and R. Ramesh, Computer-integrated process planning and scheduling (CIPPS): Intelligent support for product design, process planning and control, *Int. J. Production Research* **37**, 3 (1999) 481–500.
7. M. E. Ssemakula and A. Satsangi, Application of PDES to CAD/CAPP integration, *Computer and Industrial Engineering* **18**, 4 (1990) 435–444.
8. J. H. Vandenbrande, *Automated Recognition of Machinable Features in Solid Models*, PhD Dissertation, University of Rochester, USA, 1990.
9. J. H. Vandenbrande and A. G. Requicha, Spatial reasoning for the automatic recognition of machinable features in solid models, *IEEE Trans. Pattern Analysis and Machine Intelligence* **15**, 12 (1993) 1269–1285.
10. C. Y. Kao, *Geometric Reasoning Using Super Relation Gruph Method for Manufacturing Feature Recognition*, Master Thesis, The Pennsylvania State University, USA, 1992.
11. S. B. Joshi and T. C. Chang, CAD interface for automated process planning, *Proc. 19th CIRP Int. Seminar on Manufacturing System*, The Pennsylvania State University (1987) 39–45.
12. M. D. Gallagher, *Computational Implementation of Super Relation Graph Method for Interactive Feature Recognition*, Master Thesis, the Pennsylvania State University, USA, 1994.
13. O. W. Salomons, F. Houten and Kals, Review of research in feature-based design, *J. Manufacturing System* **12**, 2 (1990) 113–132.
14. H. T. Amrine, J. A. Ritchey, C. L. Moodie and J. F. Kmec, *Manufacturing Organization and Management* (Englewood Cliffs, NJ: Prentice-Hall, Inc., 1993).
15. H. Zhang and M. E. Merchant, IPPM-A prototype to integrate process planning and job shop scheduling functions, *Annals of the CIRP* **42**, 1 (1993) 513–518.
16. I. Ham and S. C. Lu, Computer-aided process planning: The present and the future, *CIRP Annals* **37**, 2 (1988) 591–601.
17. C. H. Link, CAPP-CAM-I automated process planning system, *Proc. 13th Numerical Control Society Annual Meeting and Technical Conference*, Cincinnati, OH (1976) 401–463.
18. J. Tulkoff, Process Planning in the Computer Age, *Machine and Tool Blue Book* 1981.
19. C. F. Sack, Jr., Computer managed process plaaing — A bridge between CAD and CAM, *The CASA/SME Autofact Conference* (1982).
20. T. W. Liao, E. R. Coates, F. Aghazadeh, L. Mann and N. Guha, Modification of CAPP systems for CAPP/scheduling integration, *Computers and Industrial Engineering* **26**, 3 (1994) 451–463.
21. K. Iwata, and Y. Fukuda, KAPPS: Know-how knowledge assisted production planning system in the machining shop, *19th CIRP Int. Seminar on Manufacturing Systems*, Pennsylvania State, USA (1987) 287–294.
22. J. P. Tsang, *The Propel Process Planner*, Pennsylvania State, USA (1987) 71–77.
23. H. P. Wang and R. A. Wysk, TURBO-CAPP: A knowledge-based computer aided process planning, *19th CIRP Int. Seminar on Manufacturing Systems*, Pennsylvania State, USA (1987) 161–167.

24. K. A. Aldakhilallah and R. Ramesh, Recognition of minimal feature covers of prismatic objects: A prelude to automated process planning, *Int. J. Production Research* **35**, 3 (1997) 635–650.
25. K. A. Aldakhilallah and R. Ramesh, An integrated framework for automated process planning: design and analysis, *Int. J. Production Research* **36**, 4 (1998) 939–956.
26. L. Alting and H. Zhang, Computer aided process planning: The state-of-the art survey, *Int. J. Production Research* **27**, 4 (1989) 553–585.
27. American Machinist, *Computers in Manufacturing* (McGraw-Hill, Inc., 1983).
28. A. B. Badiru, *Expert Systems: Application in Engineering and Manufacturing* (Englewood Cliffs, NJ: Prentice-Hall, Inc., 1992).
29. D. Bedworth, M. Henderson and P. Wolfe, *Computer-Integrated Design and Manufacturing* (McGraw-Hill, Inc., 1991).
30. T. Gupta, An expert system approach in process planning: Current development and its future, *Computer and Industrial Engineering* **18**, 1 (1990) 69–80.
31. S. Irani, H. Koo and S. Raman, Feature-based operation sequence generation in CAPP, *Int. J. Production Research* **33**, 1 (1995) 17–39.
32. R. K. Li, C. Y. Lin and H. H. Wu, Feature modification framework for feature-based design systems, *Int. J. Production Research* **33**, 2 (1995) 549–563.
33. M. E. Merchant, CAPP in CIM-integration and future trends. Pennsylvania State, USA (1987) 1–3.
34. I. Opas, F. Kanerva and M. Mäntylä, Automative process plan generation in an operative process planning system, *Int. J. Production Research* **32**, 6 (1994) 1347–1363.
35. D. Perng and C. Cheng, Feature-based process plan generation from 3D DSG inputs, *Computers and Industrial Engineering* **26**, 3 (1994) 423–435.
36. F. O. Rasch, IPROS-A variant process planning system, *19th CIRP Int. Seminar on Manufacturing Systems*, Pennsylvania State, USA (1987) 157–160.
37. S. Russell and P. Norving, *Artificial Intelligence: A Modern Approach* (Upper Saddle River, NJ: Prentice-Hall, Inc., 1995).
38. H. J. Steudel, Computer-aided process planning: Past, present, and future, *Int. J. Production Research* **22**, 2 (1984) 253–266.
39. V. Tipnis, Computer-aided process planning: A critique of research and implementation, Pennsylvania State, USA (1987) 295–300.
40. D. Veeramani, J. Bernardo, C. Chung and Y. Gupta, Computer-integrated manufacturing: A taxonomy of integration and research issues, *Production and Operations Management* **4**, 4 (1995) 360–380.
41. H. P. Wang and R. A. Wysk, A knowledge-based approach for automated process planning, *Int. J. Production Research* **26**, 6 (1988) 999–1014.

CHAPTER 5

ON-LINE REAL TIME COMPUTER TECHNIQUES FOR MACHINE TOOL WEAR IN MANUFACTURING SYSTEMS

R. J. KUO

Department of Industrial Engineering,
National Taipei University of Technology,
Taipei, Taiwan 106, ROC
E-mail: f10868@ntut.edu.tw

A critical part of a machining system in an unmanned factory is the ability to change the tools automatically due to wear or damage. In order to fit this requirement, on-line real time monitoring of tool wear becomes essential. Thus, this paper is dedicated to introduce some new computer techniques, e.g. artificial neural networks (ANNs), fuzzy logic, and methods, e.g. multi-sensor integration, for monitoring tool wear particularly in turning operations.

Keywords: Tool wear; on-line monitoring; manufacturing systems; multi-sensor integration; artificial neural networks; fuzzy logic.

1. Introduction

Tool wear is an inevitable result of the metal cutting process. Since undesirable effects of tool wear include: (1) a loss in the dimensional accuracy of the finished product (Fig. 1) and (2) possible damages to the workpiece, the on-line prediction of cutting tool wear becomes crucial. To date, it remains as one of the major obstacles to the optimization of the metal cutting process and the full implementation of unmanned machining. It is especially important for precision flexible manufacturing systems (PFMS).

There has been some research on tool wear monitoring including analytical and empirical models. Most of them are still under experiment. Recently, due to fast improvements of computer techniques, real time monitoring becomes one of the candidates for tool wear monitoring. They usually employ sensors, e.g. force, vibration, or acoustic emission, to monitor the stage of the tool wear in real time. These researches can be divided into two types: (1) single sensor and (2) multiple sensors. Using a single sensor to monitor tool wear is generally not very reliable in practice, while the multi-sensor integration method which can combine multiple sensor signals for reliable predictions and can also detect a defective sensor and then compensate for it, is more promising. Therefore, the first objective of this paper is to

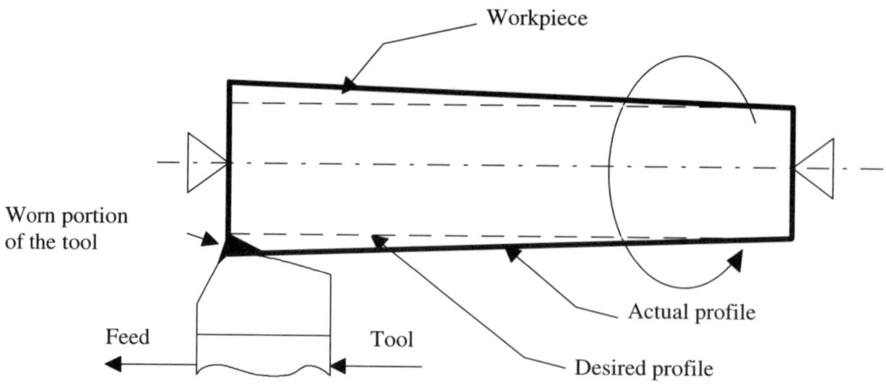

Fig. 1. The effect of tool wear on dimensional accuracy of workpiece.

introduce both the single- and the multi-sensor monitoring for machine tool wear. Besides, recent research emphasizing the application of artificial neural networks (ANNs) and fuzzy model for multi-sensor integration has been performed. Thus, the second objective of this paper is to discuss how to employ these new computer techniques for tool wear monitoring.

The rest of this paper is organized as follows. In Sec. 2, the general background of tool wear is introduced. The computer techniques applied in tool wear monitoring are discussed in Sec. 3. Section 4 presents the single-sensor tool wear monitoring, while the multi-sensor tool wear monitoring is discussed in Sec. 5. Finally, Sec. 6 presents the concluding remarks and directions for future studies.

2. Tool Wear

In metal cutting, most tools fail either by fracture or by gradual wear. Even within these two broad modes of failure, there are various other types of wear. Fracture occurs more readily in brittle tools under interrupted cutting conditions. Sometimes the fracture does not cause a complete tool failure but a small chipping of the cutting edge. During gradual wear, the tool will reach its lifespan limit by either flank wear or crater wear. There is also the depth of cut notch wear that occurs at both cutting edges in single-point machining. Generally, flank wear and crater wear, which are shown in Fig. 2, are the two most studied tool wear regions.[1]

3. Computational Intelligence

The computer technique applied for tool wear monitoring is computational intelligence. Thus, artificial neural networks (ANNs) and fuzzy set theory will be explained in the following subsections.

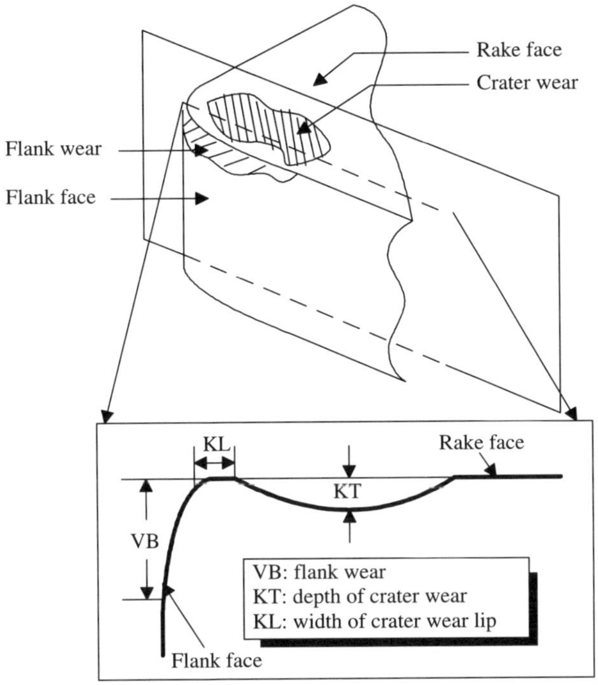

Fig. 2. Flank wear and crater wear.

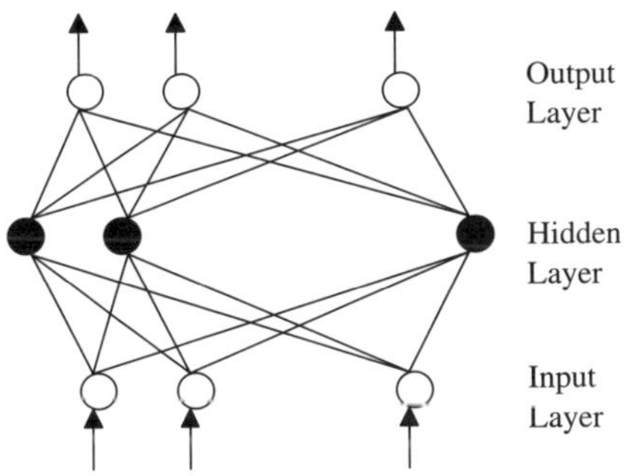

Fig. 3. Artificial neural network (ANN).

3.1. *Artificial neural networks (ANNs)*

Artificial neural network (ANN) is a system which has been derived through models of neurophysiology. In general, it consists of a collection of simple nonlinear computing elements whose inputs and outputs are tied together to form a network. The general ANN structure is shown in Fig. 3.

Generally, the learning algorithms of ANNs can be divided into three different types: supervised, unsupervised, and hybrid learning. These three learning rules are discussed more detailed as follows.[2]

Supervised learning

In supervised learning, which is always used for feedforward ANNs, the network has its output compared with the known correct answer and receives feedback about any errors. This is sometimes called learning with a teacher; the teacher tells the network what is the correct answer.

It is usually considered that the ANN includes separate inputs and outputs and it is assumed that a list of training pairs, which consists of inputs and outputs, exist. Then the connection strengths can be changed by means of minimizing the error. This is typically done incrementally, making small adjustments in response to each training pair. Among the different supervised learning algorithms, the error backpropagation (EBP) learning algorithm is one of the most studied and applied methods.

Unsupervised learning

For supervised learning, both inputs and outputs are necessary for training the network, while the unsupervised learning only needs the inputs. The network must discover for itself patterns, features, and correlations in the input data and then code for them in the output. The units and connections must thus display some degree of self-organization. The most widely used unsupervised learning scheme is Kohonen's feature maps.[3,4]

Hybrid learning

The above two mentioned learning schemes can be combined in the same network. The most common idea is to have one layer that learns in an unsupervised way, followed by one or more layers trained by the EBP learning algorithm. The reason for using hybrid learning is its good training performance. The networks have been proposed by Hecht-Nielsen[5] and Huang and Lippmann.[6] They have been called counter-propagation networks and hierarchical feature mapping classifiers. Another example of a hybrid network was examined by Moody and Darden.[7] The hidden units in the Moody–Darden network are Gaussian activation functions. Each hidden unit has its own receptive field in the input space. The Gaussians are a particular example of radial basis functions.

The reason why ANNs are so important is that the current technology has run into bottlenecks-sequential processing. When a computer can handle information only one small piece at a time, there are limits as to how fast large amounts of information can be processed. Therefore, an ANN is also called parallel distributed processing (PDP). ANNs have been shown to have the potential for solving today's technological problems, such as pattern recognition, speech/image understanding, sensor processing robotic controls, learning, etc.

3.2. *Fuzzy set theory*

The theory of fuzzy sets was first founded by Lotfi Zadeh,[8] primarily in the context of his interest in the analysis of complex systems. However, some of the key ideas of the theory were envisioned by Max Black,[9] a philosopher, almost 30 years prior to Zadeh's seminal paper. Basically, the concept of the fuzzy set is a generalization of the classical or crisp set.

The crisp set is defined in such a way as to dichotomize the individuals in some given universe of discourse into two groups: members (those that certainly belong in the set) and non-members (those that certainly do not). A sharp, unambiguous distinction exists between the members and non-members of the class or category represented by the crisp set. However, many of the collections and categories do not display this characteristic. Instead, their boundaries seem vague, and the transition from member to non-member appears gradual rather than abrupt. Thus, the fuzzy set introduces vagueness by eliminating the sharp boundary dividing members of the class from non-members.

Fuzzy sets

A fuzzy set F in a universe of discourse U is characterized by a membership function μ_F which takes values in the interval $[0,1]$ namely, $\mu_F : U \to [0,1]$. A fuzzy set F in U can be represented as a set of ordered pairs of a generic element u and its grade of membership function: $F = \{(u, \mu_F(u)|u \in U\}$. When U is continuous, a fuzzy set F can be written as:

$$F = \int_i \frac{\mu_F(u_i)}{u_i} . \tag{1}$$

When U is discrete, a fuzzy set F is represented as:

$$F = \sum_{i=1}^{i=n} \frac{\mu_F(u_i)}{u_i} , \tag{2}$$

Each membership function represents a linguistic variable. A linguistic variable can be regarded either as a variable whose value is a fuzzy number or as a variable whose values are defined in linguistic terms. A linguistic variable is characterized by a quintuple (x, $T(x)$, U, G, M) where x is the name of the variable; $T(x)$ is the term set of x; U is the universe of discourse; G is a syntactic rule for generating the names of values of x; and M is the meaning of the semantic rule.

Fuzzy logic

A logic based on fuzzy set theory is called fuzzy logic. In fuzzy logic, the fuzzy implication inference is based on the compositional rule of inference for approximate reasoning. Intuitively, the rule is of the form:

IF (a set of conditions) THEN (a set of consequences).

Generally, fuzzy modeling is a tool, which employs fuzzy set theory for complex systems. It has been widely applied in the area of control and shows very promising results.

4. Tool Wear Monitoring

In the unmanned factory, there exist an essential problem on how to automatically detect cutting tool degradation due to wear or other damages. In precision machining, it is critical to keep the tool/work distance constant. Thus to monitor the amount of tool wear on-line has become a very important research area for machining.

Like mentioned in the above section, flank wear and crater wear are the two most studied tool wear regions. Traditionally, tool change strategies are based on the most conservative estimate of tool life from past tool wear data. However, this approach does not allow for stochastic variations in tools and workpiece materials. Thus, sensors are needed to monitor the wear state.

Basically, tool wear sensing can be classified into two major categories: (1) direct sensing, where the actual tool wear is measured; (2) indirect sensing, where a parameter correlated with tool wear is measured.[1] Some direct tool wear sensing approaches are optical, tool/work distance, etc. Since these approaches require the machine to be stopped, they are not suitable for automated manufacturing. Thus, indirect tool wear sensing appears to be the only way for continuous monitoring in this application.

There are many indirect tool wear sensing approaches, such as cutting forces, vibration, acoustic emission, surface roughness, temperature, etc. Based on the number of sensors used, the indirect methods can be further divided into two types: (1) single sensor and (2) multiple sensors. Single-sensor tool wear monitoring will be discussed in the following subsection, while the multi-sensor tool wear monitoring can be found in the next section.

4.1. *Single-sensor monitoring*

Cutting forces

Measuring cutting forces is one of the most commonly used techniques in detecting tool wear. Generally speaking, as the tool wear increases, the cutting forces will also increase. The forces acting on the tool are an important aspect of machining.[10] It has been reported that cutting forces change as the tool wears.[1] There are three different forces in the feed, radial and main cutting directions (Fig. 4) which can be measured. However, various researchers claim different results. For instance, Lister *et al.*[11] found that the main cutting force provided the best indication of tool wear at any given time, while Tlusty *et al.*[12] showed that the feed and radial forces were influenced much more by tool wear than by the main cutting force. Thus, it may be feasible that all three orthogonal forces (feed, radial, and main cutting directions)

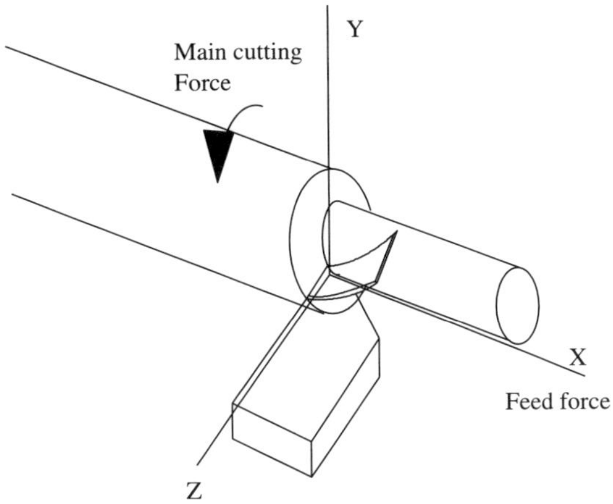

Fig. 4. Three force directions during cutting.

were monitored. Based on Rangwala,[13] and Kuo,[14] the force signals can be sampled at 1 KHz.

Vibration

Vibration results from workpiece and chips rubbing against the worn tool.[1] Thus, increased wear causes increased vibration amplitude. It can be treated as an indicator of tool wear. Basically, two accelerometers can be mounted in two orthogonal locations to measure vibration in the feed and main cutting directions. In Ref. 15 tests were sampled at 25.6 KHz.

Acoustic emission

Acoustic emission (AE) can be defined as the transient elastic energy spontaneously released in materials undergoing deformation, fracture or both. AE can be related to the grain size, dislocation density and the distribution of second-phase particles in crystalline materials and is observed during the deformation of these materials. In the metal cutting process, AE is attributable to many sources, such as the elastic and plastic deformations of both the workpiece and the cutting tool, as well as the wear and failure of the tool.[16–18] Diei and Dornfeld[19] have developed a quantitative model relating the peak value of a RMS (root mean square) AE signal to both the fractured area and the resultant cutting force at tool fracture. In the study by Rangwala,[13] the signals were sampled at 5 MHz.

Temperature

The cutting temperature increases during the cutting process, and it can affect the tool life. Technically, the final breakdown of the tool results from this increased temperature. Thus, the cutting temperature can be treated as an indicator for monitoring the tool failure. There are several techniques for assessing cutting temperatures:[1]

(1) thermo-e.f.m. measurements: work-tool thermocouple and tool thermocouples;
(2) radiation techniques; and
(3) thermo-chemical reactions.

The prime contenders are the work-tool thermocouple and the tool thermocouple techniques.

With the previous arrangements, it is difficult to measure the temperature at the cutting edge because the thermocouple is embedded within the cutting tool at a distance away from the cutting edge. This limitation was overcome using a special tool thermocouple which was utilized in the study of the relation between the cutting temperature and the tool wear in cutting glass-fiber-reinforced plastics. A small hole of 1.0–1.5 mm diameter was drilled into the workpiece and two thermocouple elements were set in this hole and fixed by bonding.[1,20]

Chow and Wright[21] developed a measurement sensor and algorithms that allow tool-chip interface temperatures to be estimated during machining. The measuring scheme relies on the signal from a standard thermocouple, located at the bottom of the tool insert, and the response time of which has been observed to be on the order of one second. The proposed scheme is an on-line measuring system. Some of the analytical models for measuring cutting temperature can be found in Ref. 22.

Surface roughness

Basically, the surface roughness of the workpiece is influenced by the sharpness of the cutting tool. Thus, surface roughness can be used to monitor the tool condition. Spirgeon and Slater[23] applied a fibre-optics transducer for the in-process indication of surface roughness during a finish turning process. Besides, surface roughness up to approximately 40 μm $R_{\max}$ can be effectively detected by applying a pair of optical reflection systems as proposed by Takeyama *et al.*[24]

4.2. *Computer techniques in single-sensor tool wear monitoring*

There already has been much research on the application of ANNs in the area of machining. Tansel[25] developed two ANN systems to represent cutting dynamics. The ANN systems are usable at any cutting speed in the 50–105 m/min range. Tansel and Laughlin[26] also used adaptive resonance theory (ART2) for the detection of tool breakage in milling operations. It provided a 97.2% success rate. Guillot and Ouafi[27] provided time domain inputs to a feedforward three-layer ANN which identified tool breakage at its output for milling. Similar applications can also be found in the research of Malakooti,[28] Khanchustambham,[29] and Elanayar.[30]

Recently, fuzzy models have also been employed in tool wear monitoring where input is always divided into several groups and having vague boundaries.

This situation is very similar to tool wear whose status is fuzzy. It has been shown how to use fuzzy models to recognize the fuzziness of tool wear status.[31] Similarly, the tool wear monitored is the tool wear state instead of the continuous value. However, most of them cannot be accurately applied in practice.

5. Multi-sensor Monitoring

5.1. *Multi-sensor integration/fusion*

The reasons that a human operator can control a system well are the use of his/her knowledge and the synergistic use of information obtained through his/her senses. Similarly, in an intelligent system, besides the knowledge base of the environment is needed, the use of information from different sensors also plays a very important role. In general, this is impossible on the basis of any single sensor domain. Hence, the system can be equipped with the different kinds of sensors in order to obtain more complete information. Therefore, in recent years, there has been a growing interest in the synergistic use of multiple sensors in order to increase the capabilities of intelligent machines or systems. Today's multi-sensor integration technology is no longer a "black box".[32] Some of the applications that can benefit from the use of multiple sensors are industrial tasks like assembly, path planning, military command, mobile robot navigation, multi-target tracking, and aircraft navigation. In other words, the objective of using multiple sensors is to provide an intelligent system to substitute for human operators. Multi-sensor integration systems need to operate in real time, they must perform integration using a variety of sensor types, and they should be effective across a wide range of operating condition or deployment environments.

There have already been some surveys on multi-sensor integration and fusion. Garvey[33] has surveyed AI approaches which can be used in multi-sensor integration and fusion. In Manns's paper,[34] he addresses some methods for high and low level multi-sensor integration based on maintaining consistent labeling of features detected in different sensor domains. Blackman[35] presented an overview of the many methods that are commonly in use in multi-sensor integration and he discussed the relative merit of these methods for future advanced systems. Meanwhile, Luo *et al.*[36] have surveyed some issues of different paradigms and methodologies of multi-sensor integration and fusion. Later, Luo and Kay[32] provided a complete survey of the increasing number and the variety of approaches to the problem of multi-sensor integration and fusion. They have appeared in literature in recent years, ranging from general paradigms, frameworks, and methods for integrating and fusing multi-sensor information, to existing multi-sensor integration methods, sensor selection strategies, and world models, along with the approaches to the integration and fusion of information from the combinations of different types of sensors. McKendall and Mintz[37] described their research in sensor integration with statistical decision theory. This paper serves as a tutorial for the analysis and the results of the specific research problems.

No matter whether internal- or external-state sensors are used, the advantage of integrating multiple sensors is that the information obtained from different sensory devices can supply accurate, global, timely, and less cost effective information.

5.2. *Multi-sensor integration for tool wear monitoring*

Rangwala and Dornfeld,[13,16,18] and Dornfeld[38,48] utilized ANNs for monitoring tool wear states in a turning operation. A multiple sensor scheme utilizing cutting force and acoustic emission information was presented. In this work, using a fast fourier transformation (FFT) yields the power spectrum representations of the time domain records. Combining the acoustic emission and cutting force spectra resulted in a vector of dimensions. Features were fed into an ANN for pattern recognition purposes. The results showed a 95% success rate for classifying binary tool wear states, fresh and worn.

Chryssolouris and Domroese[39,40] proposed an intelligent controller which uses a multi-sensor approach for process monitoring. The paper focuses on the module which integrates the sensor-based information in order to provide the controller with the best possible estimates for the tool wear and wear rate. Three techniques, ANNs, least-squares regression, and the group method of data handling (GMDH) algorithm are employed for the purpose of integration. Tests indicated that, when compared to the GMDH and least-squares regression techniques, ANNs were more effective at learning a relationship for providing parameters and estimates, especially when the relationship between the sensor-based information and the actual parameter is nonlinear. In addition, ANNs do not seem to be more sensitive (and in some cases they may be less sensitive than the other sensor integration schemes considered) to deterministic errors in the sensor-based information.

Thereafter, a statistical approach,[41,42]

$$\max \sum_{i=0}^{i=s} P_i(\theta \mid x_i), \tag{3}$$

where θ is the synthesized estimate, s is the number of sensors considered to be in agreement, x_i is the state variable estimates provided by each process model, and $P_i(\theta \mid x_i)$ is the value of the probability density function at θ given that the distribution is centered at x_i, was used. The problem of this approach is that there is no information on the probability density function of tool wear. It is typically assumed to be Gaussian. Before applying the above mentioned statistical approach for integration, the confidence distance measure for the support of sensor i by sensor j, defined as $d_{ij} = 2A$ where A is the area under the probability density curve $P_i(\theta \mid x_i)$ between x_i and x_j, is used to eliminate the non-consensus sensory values first.

Masory[43] proposed a tool wear model based on the EBP learning algorithm of ANNs. During training, the input vector to the network consists of the true RMS of the acoustic emission signal and the three components of the cutting force. Though this research tried to predict the continuous amount of tool wear, only a single cutting condition was performed. Similarly, ANNs were also applied by Kamarthi *et al.*[44] as the pattern recognizer, while the input vector was the parameter of ARMA(8, 8) and the network used was a Kohonen's feature mapping. Force and vibration sensors were used in this application. Leem and Dreyfus[45] also applied Kohonen's feature map for sensor fusion in turning. The results showed that the proposed network achieved 94% and 92% accuracy for classification into two and three levels of tool wear, respectively. Tansel[46] used ART2 to combine the information from the dynamometer and laser vibrometer in drilling. The proposed system accurately detects the pre-failure phase for all of the cases.

All the above research tries to predict the state of the wear instead of the amount of the wear except Masory.[43] In Refs. 14 and 47, Kuo and Cohen proposed an estimation system for on-line real time estimating of the amount of tool wear. The structure of the estimation system is illustrated in Fig. 5. Basically, this system can be applied to any on-line estimation system, not just the tool wear monitoring. The proposed estimation is consisted of: (1) data collection, (2) feature extraction, (3) pattern recognition, and (4) multi-sensor integration. The system first collects a sensory signal pattern, which corresponds to the particular characteristics of the process. In Fig. 5, it is assumed that there are three sensors used. From these three sensors, three sensory signal patterns can be collected. Since the pattern always contains more than 1000 data points, it is necessary to extract features which can represent this pattern. The system shows that both time series and frequency analyzer can be used. After these two analyzers extract the features from sensory signal patterns for each sensor, the features are fed into an ANN for recognition. It should be clarified that a different ANN is used for each sensor; thus, there are three total ANNs used for Fig. 5. The features from all these three sensors are not fed into a single ANN before the integration of the wear predictions from each ANN is pursued. Here, the tool wear prediction from the three different sensors are integrated by using a fuzzy model. This yields a single wear prediction.

In order to evaluate the proposed system, the 20HP LeBlonde lathe was used. Three types of sensors; force, vibration, and acoustic emission sensors, were employed (Fig. 6). For forces in the feed, radial and main cutting directions, a three-axis Kistler Z3392/b piezoelectric force dynamometer was used, while two PCB accelerometers were employed for vibrations in the feed and main cutting directions. A Physical Acoustics acoustic emission sensor, placed at the center of the tool holder, was used for monitoring acoustic emission signals. Force sensors, vibration sensors, and acoustic emission sensors were connected to the Kistler 3-channel model 5804 charge amplifier, the PCB charge amplifiers, and the DECI AE preamplifier, respectively. The force sensory outputs were connected to a National Instruments acquisition board connected to an IBM compatible PC with the LabView software package, while vibration sensory outputs were connected to a Tektronix 2630 Fourier Analyzer connected to an IBM compatible PC with a Fourier Analyzer package. The acoustic emission output was connected to an ANALOGIC acquisition board which

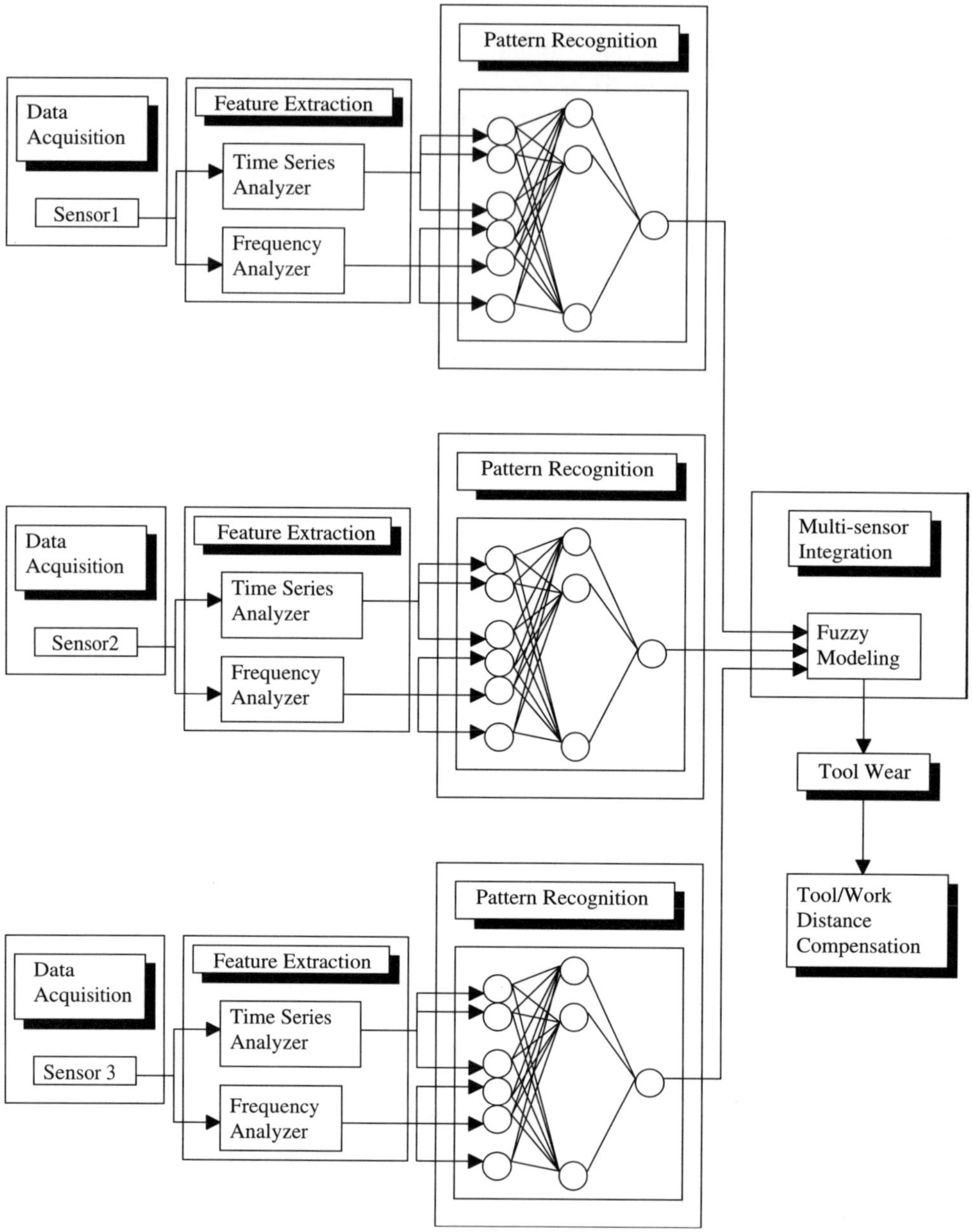

Fig. 5. The estimation system for tool wear.

was connected to an IBM compatible PC with ANALOGIC FAST Series package. The sampling rates of forces, vibrations, and acoustic emission were 3, 25.6, and 1 MHz, respectively. In addition, in order to make sure that all the three acquisition systems were triggered at the same time, an automatic trigger was connected to

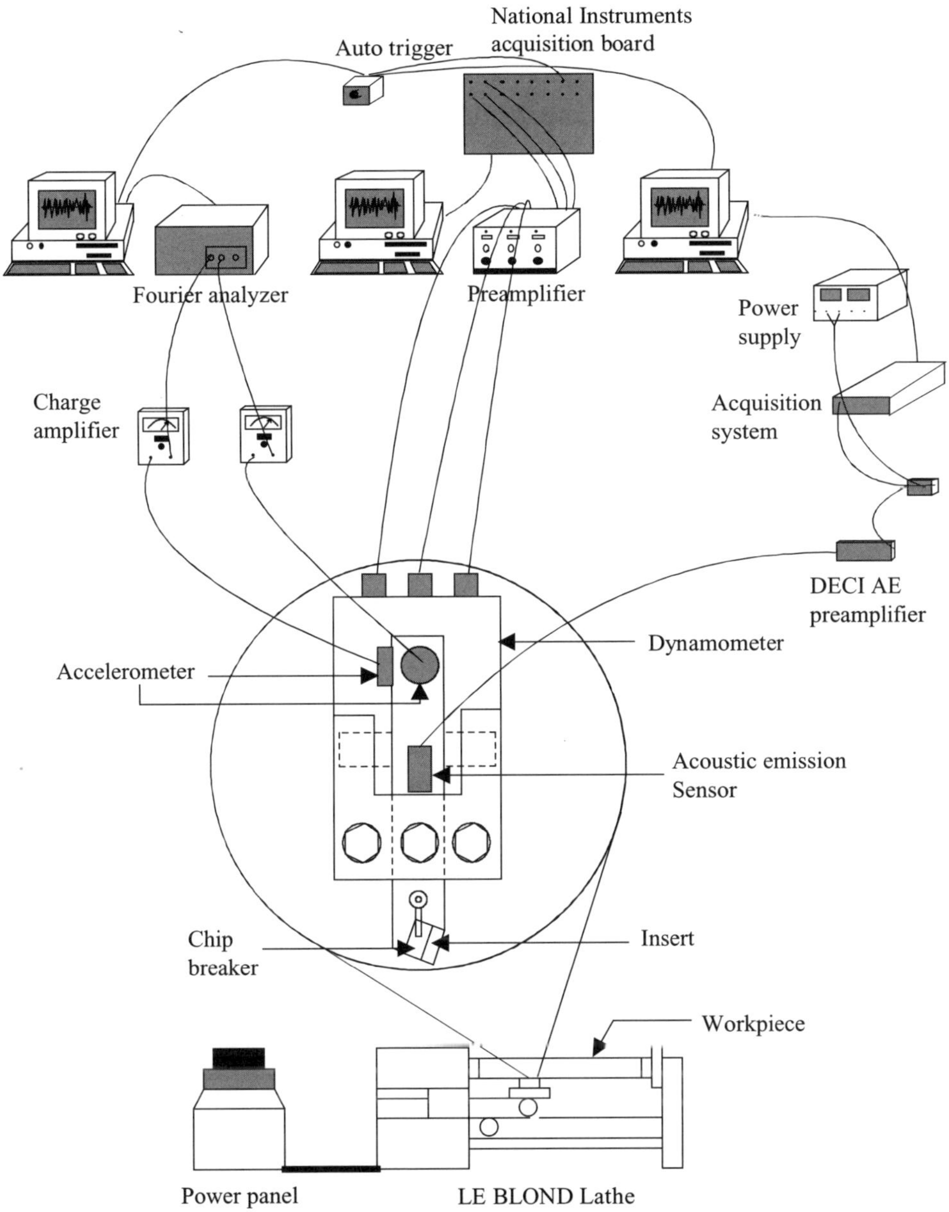

Fig. 6. The experimental setup.

the three systems. Once the trigger was initiated, it started all three data acquisition systems simultaneously. This allowed sensor data to be taken at the end of a cut and the measured wear was found to correlate with the sensor data obtained. A chip breaker was mounted on the top of the insert in order to avoid damage to

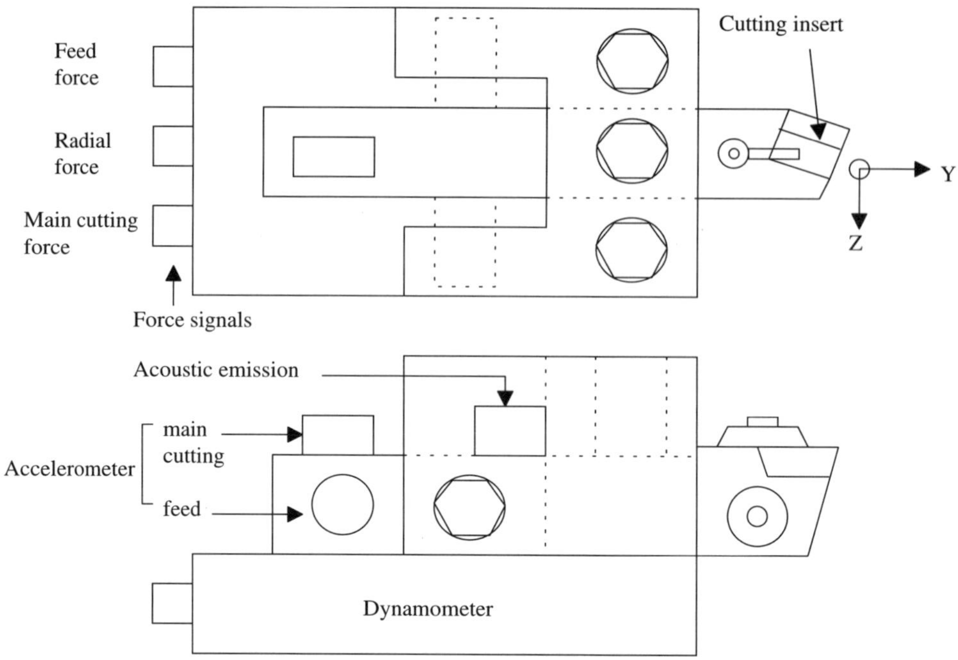

Fig. 7. The setup of sensors.

the sensors by the chips. The experimental setup and sensors setup are illustrated in Figs. 6 and 7.

The flank wear was measured by Baush & Lomb toolmaker's microscope, while surface roughness was measured using a Federal Systems Pocket Surf. A Starrett micrometer caliper was used for measuring the diameter of the workpiece.

All the sensory signals from three cutting forces in the feed, radial and main cutting directions, two vibrations in the feed and main cutting directions, and acoustic emission were collected for each cut and saved as three files. The flank wear of the tool, the diameter of workpiece, and the surface roughness was then measured off-line. All of the sensor measurements were sequenced using a common trigger just prior to the end of the cut as described.

This experiment used the SAE 6150 chromium–vanadium alloy steel as the test workpiece. The workpiece's dimensions are 7.5″ in diameter and 36″ in length. The quench and tempered heat treatment procedures of the workpiece are as follows:

(i) heated to 1550°F;
(ii) oil quenched;
(iii) tempered at 600°F; and
(iv) air cooled.

The resultant hardness ranges from 350 to 390 BHN.

The Kennamental KSBR-164C tool holder was used for machining, while the cutting insert used was a Kennamental K68 grade carbide insert SPG 422 mounted on the tool holder.

The cutting conditions were varied in order to obtained more reliable data sets. Feed rates were varied from 0.0064 ipr to 0.0156 ipr. Levels of 0.0064, 0.0088, 0.0112, 0.0136, and 0.0156 ipr were selected. Three different cutting speeds, 100, 130, and 160 sfpm, were used. The depth of cut was kept constant at 0.05 inch. Thereafter a full factorial experiment was performed. In total, fifteen different cutting conditions, or treatments (3 speeds $\times$ 5 feeds) were tried. Basically, this is the largest database collected compared with all the other related research.

The experimental procedures are described as follows:

(i) Mounting the tool insert and chip breaker on the tool holder.
(ii) Setting up the cutting conditions and calibrating the ANALOGIC FAST Series package for the acoustic emission acquisition.
(iii) Cutting the workpiece for one minute and then initiating the trigger at the end of the cut for approximately fifty-five seconds, to collect the sensory signals for the forces in three directions, the vibrations in two directions, and the acoustic emission.
(iv) Saving the sensory signals for force, vibration and acoustic emission in three different files.
(v) Removing the tool insert from the tool holder and measuring the flank wear with the Baush & Lomb toolmaker's microscope.
(vi) Measuring the diameter with the micrometer and the surface roughness using the Federal Pocket Surf.
(vii) Remounting the tool insert and chip breaker, and repeating Steps 3–6 until severe wear, about 0.018 inch, is reached.

In the third part of the estimation system, artificial neural networks are employed to recognize the features extracted from the signal patterns. Two different networks, the feedforward network with error back-propagation learning algorithm and radial basis function network, were implemented and the results for RBF network are illustrated in Fig. 8. It really provides very promising prediction.

The paper by Kuo and Cohen[14] is the first one which tries to predict the amounts of tool wear instead of the state of tool wear for multi-sensor integration. The predicted amount of tool wear can be further utilized to adjust the distance between the cutting tool and workpiece in order to obtain the more precise workpiece. The study also shows that multi-sensor integration really can improve the prediction performance as compared with the single sensor. Besides, the multi-sensor integration still can provide precise prediction in the case of sensor failure. The way to accomplish this objective is to find the inconsistent sensor and change its prediction before integration. The structure is as illustrated in Fig. 9.

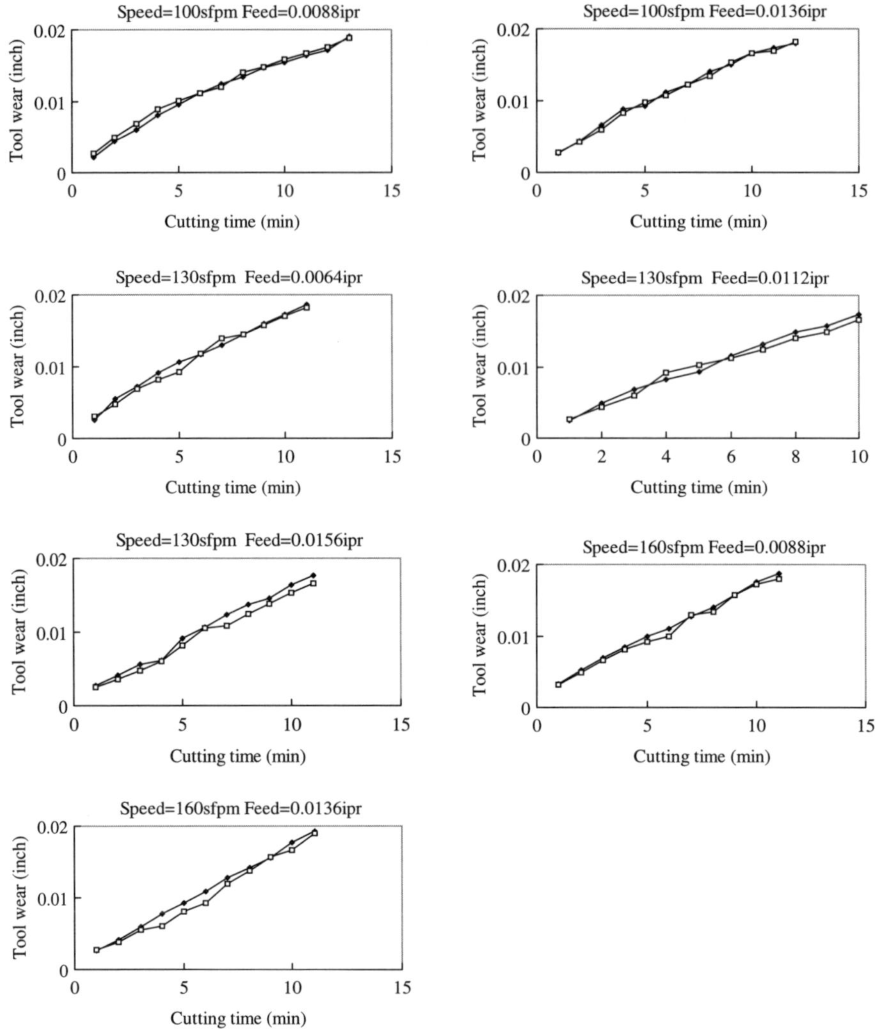

Fig. 8. Predicted (-□-) and measured (-♦-) amount of tool wear through FNN for cutting SAE6150(370BHN) with Kennamental K68/SPG422 insert.

6. Discussion and Conclusions

This paper has introduced some of computer techniques for on-line real time monitoring of the tool wear. Most of these researches are still implemented in the lab. The main reason is that the signal pattern collected from the sensor is seriously influenced by the environment. Though the computational intelligence is utilized in this area, the problem still exists. So far, on-line real time monitoring is still very difficult and it is also an open research area.

The future research can first focus on setting up the sensors. A good position, as well as protection for the sensors can make sure that the chips or the other external

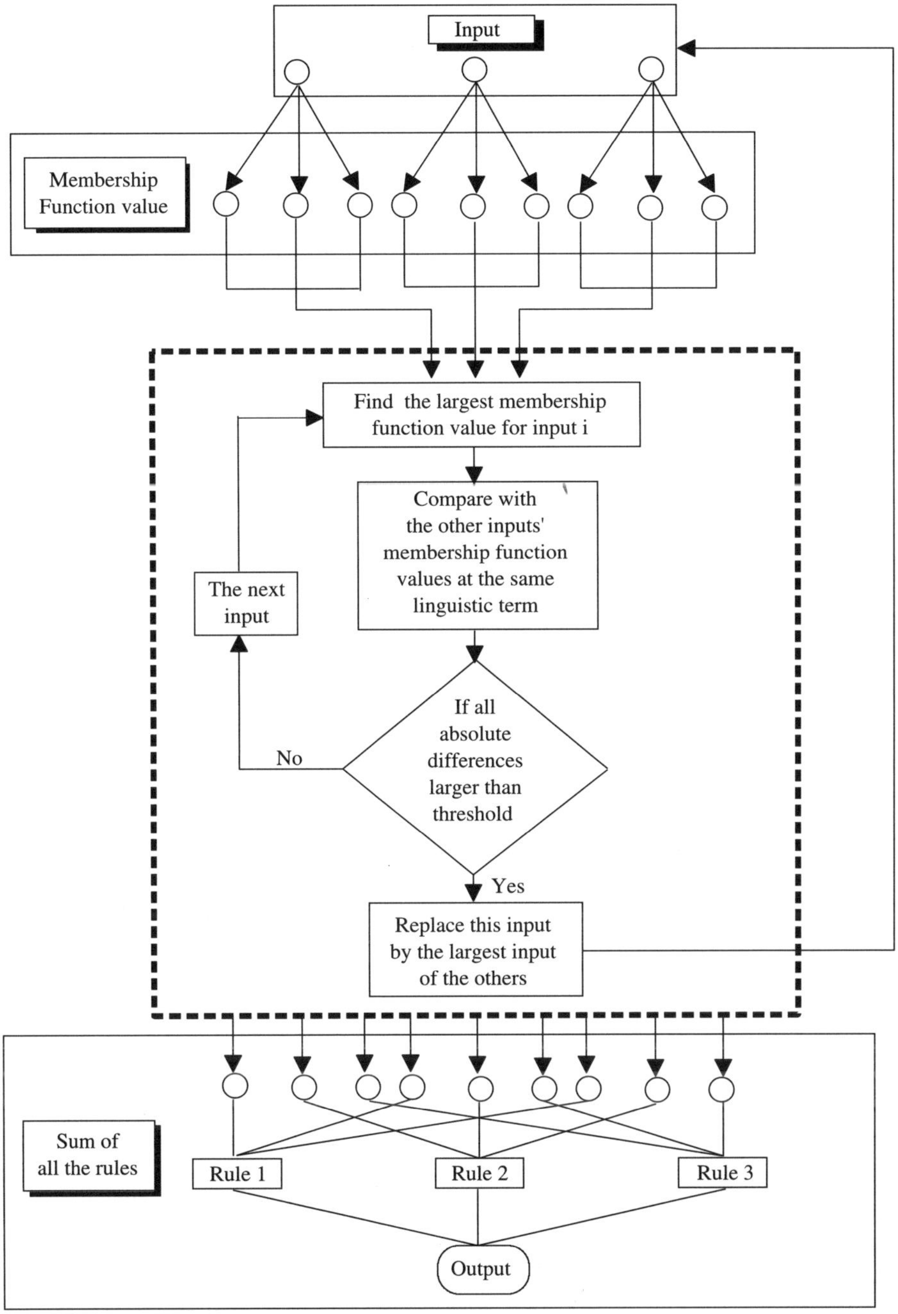

Fig. 9. The structure for detecting inconsistent sensor.

objects will not truncate the collected signal pattern. Besides, more advanced technologies of signal processing and computational intelligence can be applied. For instance, the other neural network models and multi-sensor integration schemes can be applied in the area of real time tool wear monitoring.

References

1. L. Dan and J. Mathew, Tool wear and failure monitoring techniques for Turing-A review, *Int. J. Machine Tools Manufacturing* **30**, 4 (1990) 579–598.
2. J. Hertz, A. Krogh and R. G. Palmer, *Introduction to the Theory of Neural Computation* (Addison-Wesley Publishing Company, 1991).
3. T. Kohonen, *Self-Organization and Associative Memory*, 3rd ed. (Berlin: Springer-Verlag, 1988).
4. T. Kohonen, The self-organizing map, *Proc. IEEE* **78**, 9 (1990) 1464–1480.
5. R. Hecht-Nielsen, Applications of counterpropagation networks, *Neural Networks* **1** (1988) 131–139.
6. W. Y. Huang and R. P. Lippmann, Neural net and traditional classifiers, *Neural Information Processing Systems* (1988) 387–396.
7. J. Moody and C. Darden, Fast learning in networks of locally-tuned processing units, *Neural Computation* **1** (1989) 281–294.
8. L. A. Zadeh, Fuzzy sets, *Information and Control* **8** (1965) 338–353.
9. M. Black, Vagueness: An exercise in logical analysis, *Philosophy of Science* **4** (1937) 427–455.
10. E. M. Trent, *Metal Cutting* (Butterworths & Co. Ltd., 1984).
11. P. M. Lister and G. Barrow, Tool condition monitoring systems, *Proc. 26th Int. Machine Tool Design and Research Conference* (1986) 271–288.
12. J. Tlusty and G. C. Andrews, A critical review of sensors for unmanned machining, *Annals of CIRP* 32 (1983) 563–572.
13. S. Rangwala and D. Dornfeld, Sensor integration using neural networks for intelligent tool condition monitoring, *J. Engineering for Industry* **112** (1990) 219–228.
14. R. J. Kuo and P. H. Cohen, Integration of artificial neural networks and fuzzy modeling for intelligent control of machining, *Fuzzy Sets and Systems* **98**, 1 (1998) 15–31.
15. R. J. Kuo and P. H. Cohen, Integration of RBF network and fuzzy neural network for tool wear estimation, *Neural Networks* **12**, 2 (1999) 355–370.
16. S. Rangwala and D. Dornfeld, Integration of sensors via neural networks for detection of tool wear states, *Proc. Winter Annual Meeting of the ASME* **PED 25** (1987) 109–120.
17. S. Rangwala, *Machining Process Characterization and Intelligent Tool Condition Monitoring Using Acoustic Emission Signal Analysis*, PhD Thesis, University of California, Berkeley, 1988.
18. S. S. Rangwala and D. D. Dornfeld, Learning and optimization of machining operations using computing abilities of neural networks, *IEEE Trans. Systems, Man, and Cybernetics* **19**, 2 (1989) 299–314.
19. E. N. Diei and D. A. Dornfeld, A model of tool fracture generated acoustic emission during machining, *Trans. ASME: J. Engineering for Industry* **109** (1987) 227–233.
20. K. Sakuma and M. Seto, Tool wear in cutting glass-fiber-reinforced-plastics (the relation between cutting temperature and tool wear, *Bull. JSME* **24** (1981) 748–755.
21. J. G. Chow and P. K. Wright, On-line estimation of tool/chip interface temperatures for a turning operation, *Trans. ASME: J. Engineering for Industry* **110** (1988) 56–64.

22. E. Usui, T. Shirakashi and T. Kitagawa, Analytical prediction of three dimensional cutting process: Part 3, *Trans. ASME* **100** (1978) 236–243.
23. D. Spirgeon and R. A. C. Slater, In-process indication of surface roughness using a fibre-optics transducer, *Proc. 15th Int. Machine Tool Design and Research Conference* (1974) 339–347.
24. H. Takeyman, H. Sekiguchi, R. Murata and H. Matsuzaki, In-process detection of surface roughness in machining, *Annals of CIRP* **25** (1976) 467–471.
25. I. N. Tansel, Neural network approach for representation and simulation of 3D-cutting dynamics, *Trans. NAMRI/SME* (1990) 193–200.
26. I. N. Tansel and C. M. Laughlin, On-line monitoring of tool breakage with unsupervised neural networks, *Trans. NAMRI/SME* (1991) 364–370.
27. M. Guillot and A. E. Ouafi, On-line identification of tool breakage in metal cutting processes by use of neural networks, *Proc. ANNIE'91* (1991) 701–709.
28. B. Malakooti and Y. Zhou, An applications of adaptive neural networks for an in-process monitoring and supervising system, *Proc. IJCNN'92* (1992) II-534–II-539.
29. R. G. Khanchustambham and G. M. Zhang, A neural network approach to on-line monitoring of a turning process, *Proc. IJCNN'92* (1992) II-889–II-894.
30. S. Elanayar and Y. C. Shin, Robust tool wear estimation via radial basis function neural networks, *Proc. Winter Annual Meeting of the American Society of Mechanical Engineers* (1992) 37–47.
31. T. J. Ko and D. W. Cho, Tool wear monitoring in diamond turning by fuzzy pattern recognition, *ASME J. Engineering for Industry* **116** (1994) 225–232.
32. R. C. Luo and M. G. Kay, Multisensor integration and fusion in intelligent systems, *IEEE Trans. Systems, Man, and Cybernetics* **19**, 5 (1989) 901–931.
33. T. D. Garvey, A survey of AI approaches to the integration of information, *Proc. SPIE: Infrared Sensors and Sensor Fusion* **782** (1987) 68–82.
34. R. C. Mann, Multi-sensor integration using concurrent computing, *Proc. SPIE: Infrared Sensors and Sensor Fusion* **782** (1987) 83–90.
35. S. S. Blackman, Theoretical approaches to data association, *Proc. SPIE: Sensor Fusion* **931** (1988) 60–65.
36. R. C. Luo, M.-H. Lin and R. S. Scherp, Dynamic multi-sensor data fusion system for intelligent robotics, *IEEE J. Robotics and Automation* **4**, 4 (1988) 386–395.
37. R. McKendall and M. Mintz, Using robust statistics for sensor fusion, *Proc. SPIE: Sensor Fusion III: 3D Perception and Recognition* **1383** (1990) 547–565.
38. D. A. Dornfeld, Neural network sensor fusion for tool condition monitoring, *Annals of the CIRP* **39** (1990).
39. G. Chryssolouris and M. Domroese, Sensor integration for tool wear estimation in machining, *Proc. Winter Annual Meeting of the ASME, Symp. Sensors and Controls for Manufacturing* (1988) 115–123.
40. G. Chryssolouris and M. Domroese, An experimental study of strategies for integrating sensor information in machining, *Annals of the CIRP* **38** (1989) 425–428.
41. G. Chryssolouris, M. Domroese and P. Beaulieu, A statistical approach to sensor synthesis, *Trans. North American Manufacturing Research Institution of SME* (1991) 333–337.
42. G. Chryssolouris, M. Domroese and P. Beaulieu, Sensor synthesis for control of manufacturing processes, *ASME J. Engineering for Industry* **114** (1992) 158–174.
43. O. Masory, Detection of tool wear using multi-sensor readings defined by artificial neural network, *Proc. SPIE: Applications of Artificial Neural Networks II* **1469** (1991) 515–520.

44. S. V. Kamarthi, G. S. Sankar, P. H. Cohen and S. R. T. Kumara, On-line tool wear monitoring using a kohonen's feature map, *Proc. ANNIE'91* (1991) 639–644.
45. C. S. Leem and S. E. Dreyfus, Learning input feature selection for sensor fusion in tool wear monitoring, *Proc. ANNIE'92* (1992) 815–820.
46. I. N. Tansel, Identification of the pre-failure phase in microdrilling operations with multiple sensors, *Proc. Winter Annual Meeting of the American Society of Mechanical Engineers* (1992) 23–36.
47. R. J. Kuo and P. H. Cohen, Intelligent tool wear monitoring through artificial neural networks and fuzzy modelling, *Artificial Intelligence in Engineering* **12**, 3 (1998) 229–242.
48. D. A. Dornfeld and E. Kannatey-Asibu, Acoustic emission during orthogonal metal cutting, *Int. J. Mechanical Science* **22**, 5B (1980) 285–296.

CHAPTER 6

INTERNET-BASED MANUFACTURING SYSTEMS: TECHNIQUES AND APPLICATIONS

HENRY LAU
Department of Manufacture Engineering,
The Hong Kong Polytechnic University,
Hunghom, Hong Kong
Email: mfhenry@inet.polyu.edu.hk

Recent years have seen dramatic advances in communication and information technology. These technological innovations, together with the intensified global competition, have subsequently triggered the worldwide restructuring of the manufacturing sector, causing a fundamental shift of paradigm from mass production to one that is based on fast-responsiveness and flexibility. A new pattern of production mode is on the horizon. There is no doubt that the Internet has become the world wide information platform for the exchange of all kinds of information. Intranet, which is based on Internet technology but being used within an organization, has also become a popular platform for data sharing. The advances in Intranet/Internet technology have significantly influenced the way activities are carried out among the manufacturing systems. As such, the proper deployment of this technology in the value chain of production is an essential issue to be addressed. This chapter describes the techniques and applications of Intranet/Internet technology that can be used to improve the operations among manufacturing systems.

Keywords: Fuzzy logic; Internet technology; Intranet; virtual agent; rule-based.

1. Introduction

Recent years have seen significant changes made in terms of manufacturing paradigms particularly for those companies which strive to remain world-competitive in the ever-changing market. As the manufacturing industry is becoming more unbounded by national borders, a number of global manufacturing networks have been established, taking advantage of the quickly evolving networking and information technologies. The Internet has become the common platform for the sharing of different kinds of information accessible by computers. This seems to be an irreversible trend as the deployment of Internet connection has been continually growing at an exponential rate. However, not all the people are aware that the Intranet, which is the deployment of Internet technology within a company based on open web technology, is also growing in an exponential rate. The underlying reasons why the Intranet is so well received by the public include: (i) Intranets are much easier to set up and expand and they require minimum training; (ii) Intranets can be implemented quickly. They are standards-based, they have broad vendor support and a range of product offerings; and (iii) Intranets integrate electronically to corporate data stored in servers such as product data, cost data, sales and marketing data.

The increasingly acceptance of Intranet/Internet poses a far-reaching impact on the future paradigm of manufacturing companies which are keen to innovate their manufacturing systems to enhance their ability in confronting the imminent threat of global competition. Firstly, enterprises can deploy Intranet connections within their organization and then expand it to global connection via the Internet with the objective for global information interconnection. With the implementation of Intranet within an organization, the change of manufacturing paradigm from the traditional type of sequential, function-oriented process to the simultaneous and integrated approach in dealing with operational tasks such as product development is realized.

A manufacturing organization can be divided into three main manufacturing system groups namely: marketing/sales; design; manufacturing and distribution (Fig. 1). Each group may have its own server to store the relevant information. The reality is that the company's corporate data, which includes customers data, sales and product data, is normally stored in the company's main-frame system. However, it is likely that most staff are using PCs and/or Macintoshes to run DOS/Windows/Apples applications; and due to the differences in formats and standards of the involved computer systems, it is difficult, if not impossible, to access data from the enterprise server. It should be noted that the pre-requisite of implementing the Intranet is that the corporate data must be transferred to the Web server with standards supported by Internet. Once that is done, most of the data from the server can be acccssed by clients using Web browsers as the front-end client interface — a client/server system. With this setup, various manufacturing groups can transfer information among themselves via a common interface. In addition, data communication with other groups such as design and marketing can also be achieved as most of the data is now stored in a Web server which can be accessed company-wide. Apart from internal data communication, the Intranet can easily combine with an Internet connection so that information can be transferred with local or global customers and suppliers. In general, the Intranet is "fenced off" from the external Internet by "firewalls" that allows employees to look out but withhold security control of the data that is not supposed to be accessed by outsiders.

2. Techniques and Applications

There are various techniques that can be adopted to facilitate the realization of Internet-based manufacturing systems. In this section, the techniques to be discussed include the use of virtual agents and fuzzy logic principles. In addition,

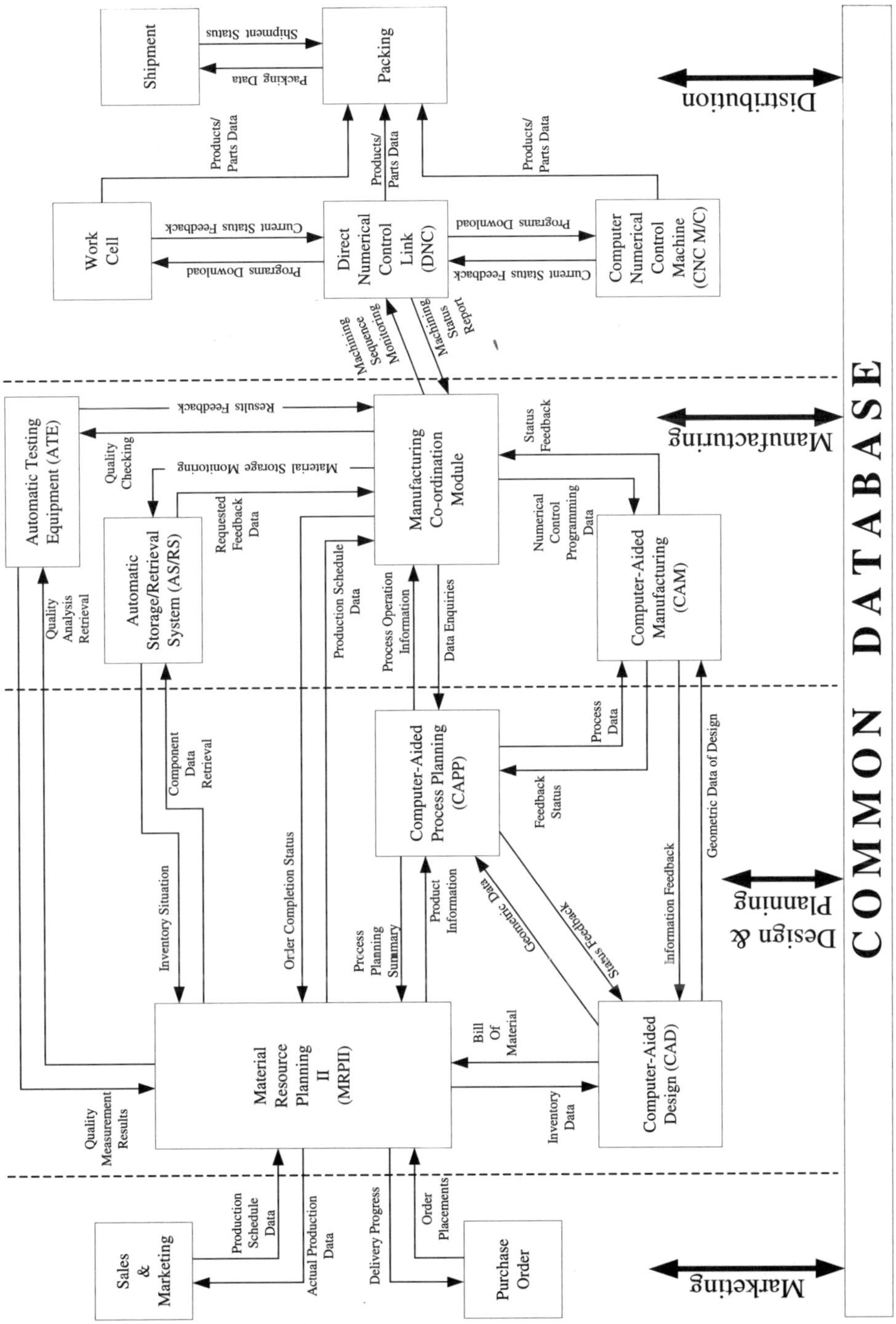

Fig. 1. Information exchange among manufacturing systems.

the applications of these techniques in enhancing the operations of Internet-based manufacturing systems are also described.

2.1. *Virtual agent techniques*

In recent years, the research related to agent-based systems which incorporate "virtual" agents for providing services to clients emulating the work of human beings, has achieved promising results in terms of the "intelligence" level of collaborative and autonomous features of agents.[1–5] These agent-based systems can be deployed in an Internet network in order to help distributed users to search for information, answer the queries of clients, communicate with novice users to solve their problems and check the security access level of users to determine the kind of information they can each view and change. In this section, an agent-based system (ABS) with the incorporation of rule-basing reasoning mechanism and object technology is described.

In general, this ABS comprises of a rule-based inference mechanism (RIM) which is responsible for the division of a client's job request into basic tasks, as well as a module comprising of a team of virtual agents for achieving automatic tasks decomposition and assignment. The agents are basically "objects" created by means of object technology.[6,7] The following shows the operations of RIM.

2.1.1. *Rule-based inference mechanism (RIM)*

An inference mechanism can be regarded, in short, as a searching process which ploughs through the knowledge base, containing facts, rules and templates, to arrive at decisions (goals). The inference process operates by selecting the rules, matching the symbols of facts and then "firing" the rules to establish new facts.[8] The process is iterated until a specified goal is arrived at.

A template in the RIM module is analogous to a structure definition of a "user-defined variable" in programming languages such as Pascal and C. For example, the template "goal-is-to" contains two "symbols", namely, *action* and *argument.* The templates are used in writing rules, the patterns of which have a well-defined structure. A template contains slots (attributes), which are either single slots or multi-slots. A single-slot (or simply, slot) contains exactly one field while a multi-slot contains one or more fields. The design of templates, facts and rules is separately elaborated below.

2.1.2. *Design of templates*

Templates are analogous to record structures in structured programming. Basically, they are designed to be used in the building of rules. In the process of decomposition of a client request, templates should be designed to suit the overall requirement particularly taking into consideration the operations of the inference process. It should be noted that although the example templates shown in the following context are designed in compliance with the specific operational process of a particular company, the same principle can be applied to the organizational processes of the other companies. The following are examples of some of the templates commonly used in companies:

(i) *On-duty-agent* — The *attributes* of the on-duty-agent include *location* (where is the agent?), *at* (the exact office-room or floor number), and *holding* (is he/she holding something or just doing something?). The pseudo-code for the *on-duty-agent* template is as follows:

Template name : *on-duty-agent*
Includes 3 attributes:
location with default value "general-building"
at with default value "common-room"
holding with default value "nothing"

This template means that the on-duty-agent has the three attributes, namely, *location*, *at* and *holding* and when the inference process of decomposition starts, the on-duty-agent is in the "common-room" of the "general-building" without "holding" anything. The code written in CLIPS[9] is as follows:

```
(deftemplate on-duty-agent
        (slot location
              (type SYMBOL)
              (default general-building))
        (slot at
              (type SYMBOL)
              (default common-room))
        (slot holding
              (type SYMBOL)
              (default nothing)))
```

(ii) *Thing* — This refers to an object which can be a dossier or an office-room. There are three attributes for the *thing* template, namely, *name* (the name of the *thing* object), *location* (where is the *thing* object?) and *at* (the exact location of it). The pseudo-code for the *thing* template is as follows:

Template name: thing
Includes 3 attributes:
name with default value "none"
location with default value "general-building"
at with default value "common-room"

This template means that the *thing* template has three attributes, namely, *name*, *location*, and *at*, and when the inference process of decomposition starts, the *thing* template has no designated name and is located in the "common-room" of the "general-building". The code written in CLIPS is as follows:

```
(deftemplate thing
    (slot name
        (type SYMBOL)
        (default ?NONE))
    (slot location
        (type SYMBOL)
        (default general-building))
    (slot at
        (type SYMBOL)
        (default common-room)))
```

(iii) *File* — This refers to a document. This template is characterized by the *unlocked-by* attribute, which means that the file has to be opened with a permit or a password. There are three attributes for the *file* template, namely, *name* (the name of the thing object), *contents* (what does it contain?) and *unlocked-by* (the permit or password required to open the file). The pseudo-code for the *file* template is as follows:

Template name: file
  Includes 3 attributes:
    name with default value "none"
    contents with default value "none"
    unlocked-by with default value "none"

This template means that the *file* has three attributes, namely, *name*, *contents*, and *unlocked-by*. So when the inference process of decomposition starts, the *file* has no designated name, it has nothing inside and it does not need to be unlocked by any key. The code written in CLIPS is as follows:

```
(deftemplate file
    (slot name
        (type SYMBOL)
        (default ?NONE))
    (slot contents
        (type SYMBOL)
        (default ?NONE))
    (slot unlocked-by
        (type SYMBOL)
        (default ?NONE)))
```

(iv) *Goal-is-to* — This refers to the goal to be satisfied. This template includes two attributes, namely *action* (the *verb* involved in the goal) and *arguments* (the *object* related to the *verb* of the goal as specified).

The pseudo-code for the *goal-is-to* template is as follows:

```
Template name: goal-is-to
  Includes 2 attributes:
      action with default value "none" which only allows one of the
              following actions: hold, unlock, change, move, on, walk-to
      arguments with default value "none"
```

This template means that the *goal-is-to* has two attributes, namely, *action* and *arguments* and when the inference process of decomposition starts, the *goal-is-to* has no designated action and argument. Notice that the attribute *arguments* is multi-slot meaning that it can contain more than one field. The code written in CLIPS is as follows:

```
(deftemplate goal-is-to
      (slot action
          (type SYMBOL)
          (allowed-symbols hold unlock change move on walk-to)
          (default ?NONE))
      (multislot arguments
          (type SYMBOL)
          (default ?NONE)))
```

2.1.3. *Design of facts*

Facts are normally asserted during the start of the inference process which operates by the selection of rules, the matching of the symbols of facts and then the "firing" of the rules to establish new facts. The assertion of facts is analogous to the initialization of a structured program, where the variables (whether user-defined variables or system variables) are assigned with certain values. In this rule-based program, the structure of the templates is used for the generation of the facts. For easy understanding, a practical example with realistic manufacturing data is adopted to illustrate the design of the facts. The facts including *on-duty-agent*, *thing*, *file* and *goal-is-to* are shown as below:

```
(on-duty-agent (location general-room) (at general-building) (holding nothing))
(thing (name general-building) (location general-room))
(thing (name doc-storage-room) (location manuf-mgr-office))
(thing (name filebox) (location manuf-mgr-office) (at doc-storage-room))
(thing (name gen-request-form) (location manuf-mgr-office) (at filebox))
(file (name gen-request-form) (contents manuf-dept-approval)
    (unlocked-by endorsement-document))
(thing (name File-Target) (location master-schedule-office) (at restricted-area))
(file (name File-Target) (contents form-for-changing-prod-schedules)
    (unlocked-by Permit-Target))
```

(thing (name document-room) (location gen-admin-office))
(thing (name permit-target-appl-doc) (location gen-admin-office)
    (at restricted-area))
(file (name permit-target-appl-doc) (contents Permit-Target)
    (unlocked-by endorsement-document))
(thing (name endorsement-document) (location prod-supervisor))
(goal-is-to (action change) (arguments form-for-changing-production-schedules))

The facts shown in the above context are self-explanatory. It should be noted that during the inference process, the *fields* of the *attributes* of the *facts* are changing continuously depending on which rules are fired. For example, the first fact, i.e. (on-duty-agent (location general-room) (at general-building) (holding nothing)), indicates that at the beginning, the agent is in the general-room of the general building without holding anything. As it will be shown in the following context, the agent will move from one place to another, holding documents and files to be authorized by relevant departments. Another point that needs to explained here is the fact "(file (name File-Target) (contents form-for-changing-prod-schedules) (unlocked-by Permit-Target))" contains the *unlocked-by* attribute. This fact means that the file-target (the "ultimate" document to be accessed for meeting the goal) contains the form for changing production schedules and it needs to be unlocked (approved for making change) by a special permit (the permit-target). The last fact is the goal of the inference process, which is to "change the production schedule of a certain production line".

2.1.4. *Formulation of rules*

Generally speaking, a rule is a collection of conditions and actions to be taken if the conditions are met. A rule is made up of two parts: the left hand side (LHS) or the antecedents consisting of a series of conditional elements to be matched against the given facts; and the right hand side (RHS) or consequents containing a list of actions to be performed when the LHS of the rule is satisfied. In CLIPS, the arrow "=>" separates the LHS from the RHS. Facts are "asserted" and modified during the inference process. In most cases and also in this example, the facts are asserted when the first rule is "fired" during the inference process.

Rules are fired in accordance with the change of the attribute field values of the goals. The example here is taken from a manufacturing firm and the service request, in this case, is to change the production schedule of a certain production line. Before that starts, the procedures required to meet this objective need to be clearly understood. To have the job accomplished, four departments are involved and all of them must agree to the change. In fact, the change of production schedule affects several relevant departments. Many related issues need to be addressed, such as the possible ramifications in the case when the goods cannot be delivered on time, and the resource problems in terms of equipment/manpower availability if the schedule is to be shortened. Firstly, the procedures required to accomplish a certain task will have to be worked out among various departments. The four departments involved in this case include the Manufacturing Manager's Office, the Production Supervisor's Office, the Master Scheduling Office and the General Administration Office. The procedures in this case include:

(i) An endorsement document has to be obtained from the Production Supervisor's Office about the request and then this form is attached with a general request form obtainable from the Manufacturing Manager's Office.
(ii) The Manufacturing Manager's Office will issue the manufacturing department approval if the request is granted.
(iii) The Manufacturing Manager's Office approves the relevant document to be sent to the General Administration Office which will check the request based on the administrative perspective and taking into consideration the reasons stated on the endorsement document (from the Production Supervisor's Office) in order to decide if a special permit for this request is to be issued.
(iv) The Master Scheduling Office considers the change approval (from the Manufacturing Manager's Office) and the scheduling situation to issue the file which together with the special permit (from the General Administration Office) will officially approve the change of production schedule as requested.

All these procedures which may somewhat differ from company to company are taken into consideration to build the rules. However, the important point is that the rules have to be "generalized" which means that they are not just designed for this particular request, as other requests of different natures should also be able to use these rules without any program rewriting.

In the rule-based expert system, inference can be done primarily in two ways, namely, forward chaining and backward chaining. Backward chaining is a goal-driven process, whereas forward chaining is data driven. As the details of these basic inference mechanisms are covered in a number of publications,[8,10] they are not described any further here. In this research, a "goal-action-field" (not goal-driven) methodology is instead introduced to design "generalized" rules used for the task decomposition mechanism. In short, the rules are grouped in compliance with the action field of the goal template and hence the name of the methodology. As illustrated in the goal-is-to template, the fields of attribute action include *unlock*, *hold*, *change*, *move*, *on*, *walk-to*, which will form the different categories of the rules.

In the group of *unlock* rules, the rules are all built based on the *unlock* field of the *action* attribute. A typical rule in this group is called "get-key-to-unlock" with pseudo-code as shown below:

The Rule with name "get-key-to-unlock"
IF The goal is to unlock a certain document for access
AND The document is stored in the common-room
AND The document has to be unlocked by a special key

AND The on-duty-agent is not holding that special key
AND The fact states that "the goal is to hold that special key" does not exist
THEN Assert "the goal is to hold that special key" as a fact in the knowledge base

In CLIPS, the code is:

```
(defrule get-key-to-unlock ""
    (goal-is-to (action unlock) (arguments ?obj))
    (thing (name ?obj) (at common-room))
    (file (name ?obj) (unlocked-by ?key))
    (on-duty-agent (holding ~?key))
    (not (goal-is-to (action hold) (arguments ?key)))
    =>
    (assert (goal-is-to (action hold) (arguments ?key))))
```

It can be shown here that the rule can deal with any document or file (a variable *?obj* in the code). The conditions are that if the document is in common-room (the default location of on-duty-agent and thing), and the document requires to be unlocked by a key (a password or any sort of authorization) which is not possessed by the agent and the goal to hold that key is non existent in the knowledge base, then the rule will be fired resulting in the assertion of a new fact which is to hold the special key. It is obvious that a number of documents have to be unlocked by special "keys". The general request form has to be "unlocked" by the endorsement document in order to obtain the approval from the Manufacturing Manager's Office.

For the group of *hold* Rules, the typical example in pseudo-code is as follows:

The Rule with name "unlock-file-to-hold-object"
IF The goal is to hold a certain document
AND A certain file (say, file-A) contains that document
AND The fact states that "the goal is to unlock file-A" does not exist
THEN Assert "the goal is unlock file-A" as a fact in the knowledge base

In CLIPS, the code is :

```
(defrule unlock-file-to-hold-object ""
    (goal-is-to (action hold) (arguments ?obj))
    (file (name ?file) (contents ?obj))
    (not (goal-is-to (action unlock) (arguments ?file)))
    =>
    (assert (goal-is-to (action unlock) (arguments ?file))))
```

It can be shown from the two typical rules in the *unlock* and *hold* rule-groups that the structure of the rules are characterized by the first line of the LHS, which indicates the goal with the emphasis on the *action* field of the *goal-is-to* template. In fact, most of the rules are designed with this "goal-action-field" methodology which categorizes the rules based on the field of the action attribute of the goal template and the consequent is another goal with probably a new action field such as *unlock* in the above example. It should be emphasized here that the rules should be "generalized", which means that it is not designed for only one type of goals. They should be able to cope with various goals as those rules are basically designed in accordance with the action field of the goal-is-to template as well as the actual operational process of the company.

2.1.5. *Object-oriented virtual agent (OOVA) module*

With the basic tasks available, the next step deals with the execution of these tasks by the responsible agents. The OOVA module unit contains the details of the virtual agents which are objects created by the object-oriented programming tool. A number of tools can be used to develop these objects. The Window-based ones include Visual Basic, Delphi, PowerBuilder, Visual C++ and others. In object-oriented programming, most of the objects contain elements such as attributes, object methods and interface with the outside world.[11,12] The detailed functions of these elements are described in most of the programming books for object-oriented programming and therefore not to be covered here.

Generally speaking, each object is responsible for performing some duty depending on the *methods* and *attributes* encompassed within the object. For example, the security agent (object) is responsible for checking the access level of users so that it can determine what sort of information the individual users can access. The sample code written in VB5 is presented below:

```
Public Sub entry_security_check()
  Dim no As Long
  Dim i As Integer
  get_number_of_obj
  For no = 0 To number_of_obj - 1
  retrieve_obj_content no
  If InStr(1, Trim(username), Trim(frmlogin.f_username)) Then
    If Trim(password) = Trim(frmlogin.f_password) Then
      frmlogin.Hide
      frmWelcome.Show 1
      obj_no_found = no
      check_interests
      Exit Sub
    End If
  End If
  Next
    MsgBox "Invalid Entry !"
End Sub
```

The name of this method is entry_security_check which is used to check the password (line 7 & 8) entered by the user against the security datafile. The *for-next* loop in the code above is used to check through the list of users from the database and then extract the system access level of the user. There are levels of security to be specified by the Administration Office. In this example, there are different access levels. At the highest level, the user can change some important data such as the updating of facts and the rules in the system database. For those without an authorized password, they can also view some data which are open to the public.

Based on the features of objects in object-oriented programming,[11,12] one object can access the methods of another object by creating an instance of the other object using the command:

Set instance_of _objectA = New objectA

The object *instance_of_objectA* is now an instance of *objectA* and can access some methods of *objectA* as long as these methods are declared to be publicly accessible. Objects can communicate and exchange information by virtue of this feature.

It should be noted that like the human agents in companies, virtual agents may also face the situation that they may be phased out or modified and in some cases other new agents may be added to the system as well. Various agents (objects) contain their own relevant methods for performing duties but the next immediate question is how to coordinate the agents to carry out the separate tasks. This issue will be dealt with in the following section.

2.1.6. *Task control subsystem (TCS)*

The TCS plays the role as a coordinator as well as the administrator for the RIM and OOVA modules. It performs two important functions: (a) monitoring the status of the basic tasks deduced from the RIM and (b) coordination of the tasks to be carried out by the relevant VAs in compliance with the type and nature of tasks to be completed.

The basic tasks produced after the decomposition process have to be monitored and assigned to the relevant VA for processing. The TCS will first check through the recommended actions deduced from the inference engine to ensure that the agents in the OOVA module are able to carry out the tasks. If any one of the tasks cannot be processed by any of the included agents, the user has to be informed of this so that an alternative solution could be worked out. When the TCS is satisfied that the tasks can be done by the included agents, commands will be sent to the relevant agents for task execution. It is important that the TCS should follow closely every process carried out by the responsible agents and to ensure that individual agents will be assigned with the tasks deduced and the whole job is not considered completed until the goal, in this case the "change of schedule form", is achieved.

In order to ensure smooth and efficient exchange of information between the RIM and the OOVA modules, it is important that they are working under the same operating environment. For example, if the RIM is developed with the expert system shell called CLIPS while the OOVA module is developed with an object-oriented programming tool VB5, these two development tools cannot "naturally" talk to each other. In this aspect, it is important that these two modules should be "integrated" in order to achieve efficient bi-directional data transfer. Fortunately in Microsoft Windows, some dynamic link library (DLL) programs can be developed to link Window-based products to achieve information exchange among the software applications. As a matter of fact, there are DLL programs available for integrating CLIPS to VB5. These programs include "clips.dll" and "clipshll.dll" which can be downloaded from the Internet.* With these DLL programs added to VB5, the inference mechanism of CLIPS becomes a part of VB5, thus enabling free and automatic data exchange between these two modules.

As the inference mechanism becomes part of the object-oriented programming environment (in this example VB5), the list of tasks generated is directly sent to the TCS which is a program within VB5. The task items are treated as the *list items* inside a *ListBox*.[11,12] The task items will be collected one at a time and the content is checked to decide which agent is responsible in carrying out which task. A command in VB5 called *Instr()* can be used to check the keywords within the "string". A function of TCS called Extract_Keywords() is used to extract the keywords of the tasks. Notice that for every task, there are "pairs" of words; one is the *movement* word and the other is the *destination* or *object* word. For example, if there are three "pairs" of movement-object keywords, namely: (i) *takes* and *Gen-request-form*; (ii) *off with* and *Filebox*; and (iii) *onto* and *Common-room*, these three pairs of keywords in this case can sufficiently suggest which agent is responsible for the relevant task. The TCS will first extract the keywords from the tasks deduced by the RIM using a specially-built function Extract_Keywords (task), the code of which (in Visual Basic) is shown below:

```
Private Sub Extract_Keywords(task)
  Dim pos As Integer
  Dim num As Integer
  Dim movement(5) As String
  Dim destination(5) As String
  movement(1) = "approaches"
  destination(1) = "common-room"
  movement(2) = "gets access through"
  destination(2) = "manuf-mgr-office"
  movement(3) = "obtains"
  destination(3) = "FileBox"
  movement(4) = "strolls off"
  destination(4) = "gen-admin-office"
```

*The web site address is: http://ourworld.compuserve.com/homepages/marktoml/clipstuf.htm.

```
    movement(5) = "opens"
    destination(5) = "file-target"
    ......
    ......
    movement(n) = "opens"
    destination(n) = "file-target"
    pos = 0
    key_words = " "
    For num = 1 To N
      pos = InStr(pos + 1, basic_task, movement(num))
      If pos Then
          key_words = key_words & movement(num) & "."
      End If

      pos = InStr(pos + 1, first_task, destination(num))
      If pos Then
          key_words = key_words & destination(num) & "."
      End If
    Next
End Sub
```

This function first declares an array with all the *movement* and *destination/object* keywords included. Then the first deduced basic task is taken and checked against the whole array using the for-next loop from 1 to N (the final element of the array). The function instr() returns the position of occurrence of the relevant keyword within the basic task. If the keywords are found, they will be concatenated by the symbol "·". For example, the "string" of keywords found for the first basic task "on-duty-agent strolls off the general-building onto the common-room" will be "strolls off.general-building.onto.common-room.". With this technique, all the keywords of the basic tasks are to be extracted and joined together by the "full-stop" sign.

The next step is to assign the task to appropriate agents based on the keyword-string. TCS invokes a specially-written subroutine called *assign_based_on_keywords()* which is designed for the assignment of tasks to appropriate agents. It should be noted that the *assign_based_on_keywords()* subroutine only suggests the agents who are considered suitable for the relevant tasks.

2.2. *Application of ABS*

This technique of using virtual agents can be deployed in the Internet for globalized manufacturing. For example, Microsoft's Visual Basic can be adopted to transform the ABS to the Internet for providing service to customers via the Web.

A software based on this technique can be developed and implemented in accordance with the principle outlined above. There are three phases as described below:

(i) The first phase involves the technical evaluation of the design and operation of the prototype software by verifying that the task decomposition and assignment subroutines are executed as expected. In particular, the integrated group of OOVA module and RIM with the embedding of CLIPS into the VB5 environment is to be tested to ensure that the automatic decomposition and assignment of tasks is successfully executed.

(ii) The second phase deals with the basic system evaluation. The prototype program with modifications made according to the problems encountered in the first phase is deployed on a local area network (LAN) in the first instance and the results are observed and recorded by the project team members. The long-term objective of Internet-based operation is left to a subsequent development. Some manufacturing data such as production schedules, manufacturing products details and product design data are created so that the situation emulates a real "small-scale" manufacturing environment. The purpose of this task is to determine whether the tested system can come up with consistent as well as correct responses with regards to the clients' inquiries.

(iii) The third phase is concerned with the overall site evaluation of the system. It is important so that the ABS is able to be linked with other subsystems and the integrated system is able to be field-tested by the real end-users in order to determine the possible problems when operating in a practical manufacturing environment. Before this evaluation, the application software has to be modified to suit the actual situation. This would require significant software updating of the original prototype program and therefore the cost involved in software coding in this phase is one of the main considerations.

The implementation demonstrates that it can be used in an actual industrial environment. This system is favorable to the progressive introduction of machine intelligence features into the operation and is able to enhance the operational efficiency of manufacturing systems particularly in the Intranet/Internet environment.

2.3. *Fuzzy logic technique for assessing customer interest*

Experience indicates that it is easier to get first-time visitors to a Web site than to keep them as repeat users. Attracting repeat customers is not an easy task. Site users will revisit a particular site when they find that the relevant Web site can effectively deliver the information they are after without even being requested. In this aspect, the automated delivery of the information likely favored by the relevant visitors is an essential factor to make them frequent the site so much as to eventually strike a business deal with the Web site owners. It would be ideal if the Web site is able to find out the preferences of visitors by studying their movements covering the various Web pages. More importantly, a Web site which can progressively "learn" the specific interest of the relevant users can significantly enhance the attractiveness of that Web site. With this detected data absorbed and learnt by the system while the visitors are navigating the site spots in accordance with their natural tendency of movement, the favorite preferences of the relevant visitors can be determined and appropriate information can be delivered to them in a timely manner. This basically eliminates the requirement of traditional Web sites which ask for specific input data before relevant information is sorted for them. With this proposed approach, visitors are able to obtain the information they are after without going through the traditional requirement of inputting data into a search engine. When they visit the same Web site again, information centered on the interest of relevant visitors is delivered. This can significantly enhance the functionality and attractiveness of the Web site thus greatly raising the competitive edge of manufacturing companies.

In general, this fuzzy logic technique can be applied to the Internet-based global manufacturing area to enhance the quality of customer service and in particular the delivery of the information favored by individual customers.

2.3.1. *Overview of Web business*

Recent advances in Internet technology offer dramatic opportunities for innovative applications in the diverse business areas. As a matter of fact, the Internet has become the world-wide information platform for the sharing of all kinds of information. To take advantage of this increasingly popular "information superhighway", a number of commercial and non-commercial organizations have been connecting their companies to the World Wide Web. The Global Manufacturing Network[13] home page created by the Society of Manufacturing Engineers provides a list of products to customers who can obtain the products from selected on-line vendors. 3M, which is a multi-national company with a diverse range of products from stickers to pharmaceutical products, has also created its own Web site named the 3M Innovation Network[14] which has incorporated some information delivery concept with a list of new products available to relevant site visitors. Other successful Web sites also includes the Montgomery Wong's Webshop[15] where clients can go window shopping around several shopping spots within the Web site to browse various items, from software to sandwiches prior to placing orders and paying bills, all done via the Web. Any visitor can register as a customer and once registered he/she will be recognized at the Web site in the following visits. This can provide some sort of "personalized" information to the customer such as the list of items bought so far, payment account record for the past months, etc.

Ching[16] points out that automatic information delivery is meant to promote the products and image of the company but not without any pitfalls. In recent years, Internet technology developers have been working along the line of the automated delivery of "personalized" up-to-date information generally coined as webcasting.[17–19] Despite the improved functionality of Web sites with the incorporation of automatic delivery concepts such as the personalization features offered by

Microsoft site server,[26] users are either likely to receive a huge amount of information not of their interest, or that the useful information is not available to them in time.

Basically, the search of information of these Web sites adopts traditional approaches including search engines and hierarchical indices list where the specific input by visitors is essential. It should be noted that some visitors may not have a specific site to visit when they navigate through the Internet. In normal cases, site visitors just want to try their luck if anything happens to be of their interest. Obviously these traditional search approaches have not fully served the purpose to provide satisfactory service to these visitors. This Internet information delivery system (IIDS) is able to automatically "learn" to assess the interests and preferences of the individual visitors by following the site movement of visitors with real time evaluation at the background. This system can be deployed in the manufacturing systems of enterprises so that the updated manufacturing information (see Fig. 1) can be shared and used "in real time", thus enabling a better chance of striking a purchase or a business deal from potential investors and customers.

2.3.2. *Principle of Internet information delivery system (IIDS)*

An Internet information delivery system is proposed in this chapter, embracing primarily the fuzzy logic principle for the evaluation of the site preference of relevant visitors. In brief, fuzzy logic principle[20] is based on a "superset" of Boolean logic that has been extended to handle the concept of "partial truth" and it replaces the role of a mathematical model with another that is built from a number of fuzzy variables such as output temperature and fuzzy terms such as "hot", "fairly cold", and "probably correct". As mentioned in the above context, the underlying technology of IIDS is the fuzzy logic principle, the detailed theory and application examples of which can be found in a number of publications[10,21–24] and therefore not to be elaborated in this chapter.

To develop IIDS, the prerequisite is to formulate the design of a fuzzy basis mathematical model to calculate the interest level indicating point (ILIP) for individual site visitors. For example, visitors can freely walk around seven shopping spots within the Cyber Mall web site. They are: business equipment (BE) shop, computer (Com) shop, food/wine (F/W) shop, interior design (ID) shop, office (Off) shop, shopping mall (SM) and telecommunication (Tele) shop.

In this IIDS, the statistical data on visiting duration, i.e. Ts (total time of stay of a visitor of a certain spot), Fv (frequency of visits of a certain spot over the total calculated period) and Ti (visiting interval) are the fuzzy input variables.[25] It is obvious that Fv and Ts are important factors to determine the level of interest of Web visitors on certain site spots. In particular, Ti can reflect a more precise and interesting level of a visitor for the visited spots based on the touring path of each visit. In brief, the visiting interval is defined as the time elapsed between any two successive visits of the same spot.

The three functions, i.e. total time of stay Ts, visiting frequency Fv and visiting interval Ti, are treated as inputs of the fuzzy basis IIDS. The membership function μ of these functions are defined as a set of ordered pairs and their membership grade values. The fuzzy sets F_{Ts}, F_{Fv} and F_{Ti} for the functions in this investigation are given as:

$$F_{Ts} = \{(Ts, \mu_{Ts}(ts)) \mid ts \in Ts\}\,, \tag{1a}$$

$$F_{Fv} = \{(Fv, \mu_{Fv}(f)) \mid f \in Fv\}\,, \text{ and} \tag{1b}$$

$$F_{Ti} = \{(Ti, \mu_{Ti}(ti)) \mid ti \in Ti\}\,. \tag{1c}$$

In Eq. (1), Ts, Fv and Ti are the universal sets of the three input functions, and ts, f and ti are the elements of the corresponding universal sets. The membership functions μ_{Ts}, μ_{Fv} and μ_{Ti} show the probability of the variables in their own set. Figure 2 illustrates the characteristics of the three membership functions and their labels.

Five triangular shape membership functions with 50% overlapping are selected as the input functions in this investigation; this setting can provide a smooth fuzzy reasoning result in comparison with non-linear membership functions sets. The linguistic labels for the membership functions of μ_{Ts} and μ_{Ti} are the same, i.e. short (S), rather short (RS), normal (N), rather long (RL) and long (L), see Figs. 2(a) and 2(c). The labels for the membership function of Fv as illustrated in Fig. 2(b) are designed as low (LO), rather low (RLO), normal (N), rather high (RHI) and high (HI). The universe of discourse for the elements ts, f and ti are fixed in a range of minimum (Min.) to maximum (Max.) values from the statistical records of each trip, and these values vary from one trip to another trip.

2.3.3. *Output functions and fuzzy rules*

Basically, the output of the IIDS is the interest level indicating point (ILIP). It is designed to indicate relative interest levels of the visited shopping spots of a trip. In the system, a variable W is assigned as the output function of the fuzzy set.

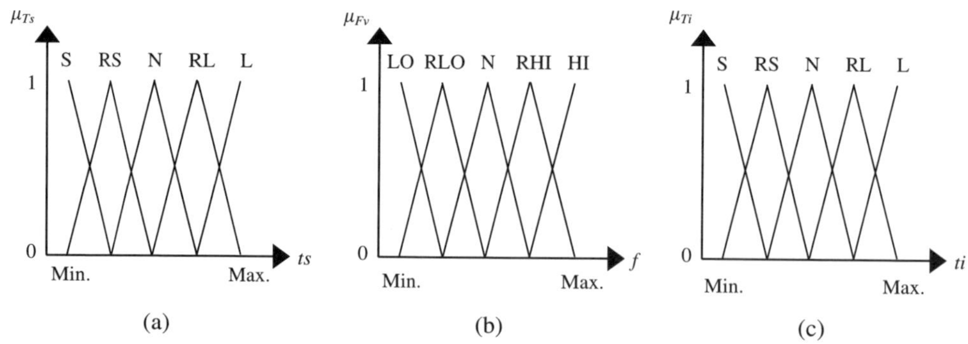

Fig. 2. The characteristics of the three membership functions.

This variable is named 'weight' and the elements w in this fuzzy set are designed in a fixed range from 0 to 100. In other words, ILIP is the defuzzified result of W. The fuzzy set of W is written as:

$$F_W = \{(W, \mu_W(w)) \mid w \in W\}\,. \tag{2}$$

A fuzzy rule base is required to set up in order to determine the results of W from the three input functions, i.e. Ts, Fv and Ti. The rules are usually connected into aggregated statements using logically operators, such as AND. The expressions of each rule are simply coded as a direct cause and result description, e.g. IF ... AND ... THEN .... For the Cyber Mall (the Web site created for the test of IIDS) as designed in this investigation, a total of 125 fuzzy rules were defined. Figure 3 shows the corresponding membership functions as labeled from #1 to #41 for the entire rule base in a 3-dimensional matrix. The matrix is constructed by overlapping five numbers of 2-dimensional matrices. Each 2D matrix is defined from one of the five membership functions of Fv as illustrated in Fig. 2(b). For each 2D matrix, two input function, i.e. Ts and Ti are used to define 25 fuzzy rules. In fact, a larger linguistic label number will be obtained for a longer Ts and shorter Ti. The labeled number also increases when the value of Fv moves up. The same rule will be repeatedly used for any two consecutive membership functions of Fv in order to provide a smooth transition between two successive 2D matrices, e.g. the membership functions of #9, #17, #25 and #33 as shown in Fig. 3. The 41 membership functions for the weight W are represented by triangular shape with 50% overlapping as illustrated in Fig. 4.

When one of the 125 rules is fixed, a linguistic term of the output function W can immediately determined from Fig. 3, and a corresponding membership function μ_W

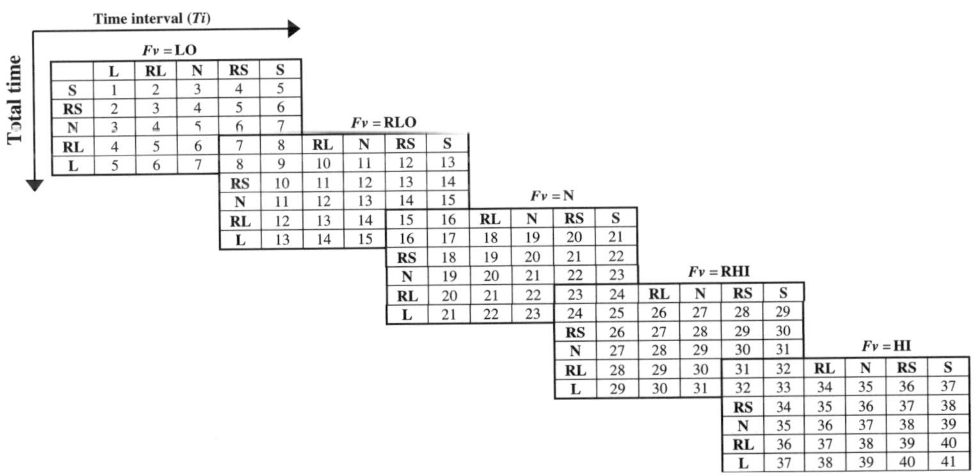

Fig. 3. Membership functions as labeled from #1 to #41 for the entire rule base in a 3-dimensional matrix.

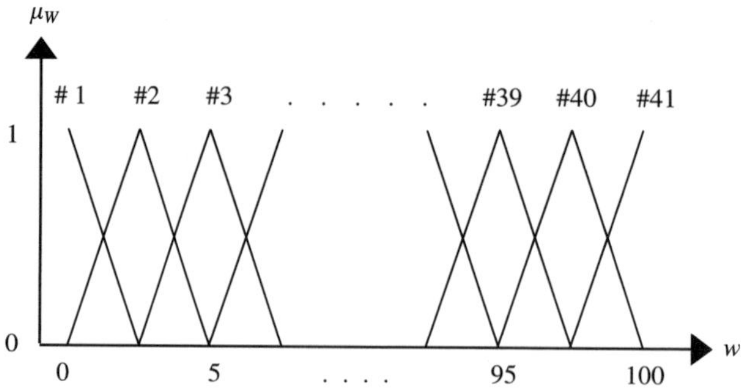

Fig. 4. The 41 membership functions for the weight W.

is also found from Fig. 4. As stated before, each fuzzy rule is defined in a direct cause (antecedent) and result (consequent) form, they can be written as:

$$\begin{array}{l}\text{IF } X_{1,1} = a_1 \text{ AND } X_{1,2} = b_1 \text{ AND } X_{1,3} = c_1 \text{ THEN } Y_1 = d_1 \\ \text{IF } X_{2,1} = a_2 \text{ AND } X_{2,2} = b_2 \text{ AND } X_{2,3} = c_2 \text{ THEN } Y_2 = d_2 \\ \quad \vdots \qquad\qquad \vdots \qquad\qquad \vdots \qquad\qquad \vdots \\ \text{IF } X_{n,1} = a_n \text{ AND } X_{n,2} = b_n \text{ AND } X_{n,3} = c_n \text{ THEN } Y_n = d_n\,. \end{array} \tag{3}$$

In this fuzzy set, X and Y represent the antecedent and consequent set functions respectively; a, b, c and d represent the linguistic labels in their own fuzzy set. In fact, smaller # implies w is closely to 0, and larger # implies w is closely to 100 as illustrated in Fig. 4.

2.3.4. *Fuzzy reasoning and defuzzification process*

Fuzzy reasoning is a process using mathematical methods to calculate the fuzzy output probability from the probability values of the input functions. For example, the probabilities for their own set of Ts, Fv and Ti in the IIDS are given as $\mu_{Ts}(ts)$, $\mu_{Fv}(f)$ and $\mu_{Ti}(ti)$ respectively. When the sub-product[21] reasoning strategy is used, the probability of fuzzy output function $\mu_W(w)$ can be written as:

$$\mu_W(w) = \mu_{Ts}(ts) \bullet \mu_{Fv}(f) \bullet \mu_{Ti}(ti). \tag{4a}$$

The interpretation of Eqs. (3) and (4a) in graphical form is illustrated in Fig. 5(a). Commonly, more than one fuzzy rule as stated in Eq. (3) will be fired in reasoning process. In other words, m numbers of linguistic label d and their corresponding $\mu_W(w)$ will be obtained when m numbers of rules are fired. The output probability functions, i.e. $\mu_W(w)$ are a group of triangles as shown in Fig. 5(b). The triangles are then used to construct a polygon as illustrated in Fig. 5(c) by superimposing methods in the sub-product reasoning strategy.

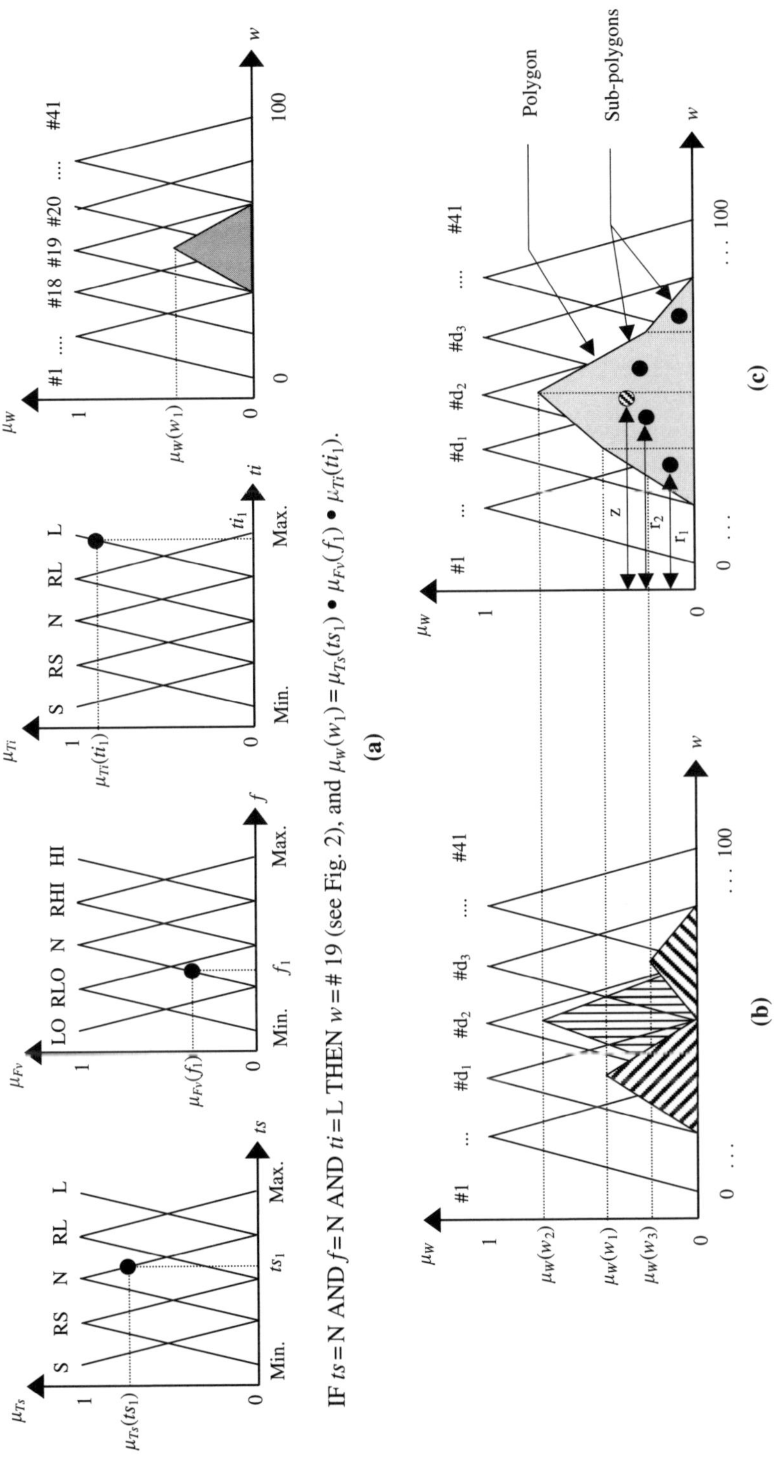

Fig. 5. The interpretation of equations in graphical form.

In mathematical expressions, the operations of superimposing m triangles is given as:

$$\mu_{Wm}(w) = \mu_W(w_1) \vee \mu_W(w_2) \vee \mu_W(w_3) \vee \cdots \vee \mu_W(w_m). \tag{4b}$$

Clearly, the operators $\bullet$ and $\vee$ in Eq. (4) represents the dot product and numerical summation respectively.

Determining a crisp output value in a non-fuzzy space from the result of fuzzy reasoning is called defuzzification. In simple words, the defuzzification process is to calculate a crisp value z between 0 to 100 for the ILIP from the polygon as constructed by the sub-product fuzzy reasoning strategy. Due to popularity, the centre-of-area (COA) method is employed in the IIDS. In this method, the centre of gravity of the polygon z is determined by finding the average gravity from a group of sub-polygons they are used to construct the polygon, see Fig. 5(c). The equation of calculating z is given as:

$$\begin{aligned} Z &= \frac{\int_l^m r_w \mu_w(w)\,dw}{\int_l^m \mu_w(w)\,dw}, \quad or \\ z &= \frac{\sum_{i=1}^m r_i A_i}{\sum_{i=1}^m A_i}, \end{aligned} \tag{5}$$

where r and A are the moment arm and the area of the sub-polygons respectively.

2.3.5. *Application of IIDS*

A trial needs to be carried out to validate the principle of the IIDS. An interactive Web site was set up. The authorized Web users can freely navigate through the seven shopping spots with unlimited number and visiting time. They also can place purchasing orders using a personal password as provided by the web master. The touring path at each visit of a visitor will be automatically recorded through a web counter as coded in Visual Basic Script.[27] Three sets of statistical data are extracted from the recorded information of the touring paths, they are the total time of stay for each visited spot Ts, the visiting frequency of each spot Fv, and the time elapsed between two successive visits of a spot Ti. An ILIP for each spot is then determined by the investigated IIDS. The visited spots with relatively higher ILIP values means that the visitor has greater interest on the spots. The web master will put more product information on these spots, and will place product promotion pages of the spots on the front page of the Web site for the next visit by the same visitor. It may even be possible that *real time* information delivery can be achieved provided that the fuzzy reasoning engine working at the background is able to operate as desired. This proposed system possesses human-like decision ability related to the sale of products through the Internet under the concept of selected information delivery and therefore it will be more successful than the traditional massive information delivery strategy. This argument is subsequently validated by a trial.

Procedure 1: Data collection and listing
The touring information from three different site visitors was randomly extracted from the database of the IIDS and listed in Table 1. The required statistical data for the calculation of ILIP were worked out from the information in Table 1 and listed in Table 2. They are the Ts, Fv and Ti.

Procedure 2: Fuzzy reasoning
The statistical data as shown in Table 2 are used to set up fuzzy sets for the three fuzzy input functions. Due to the reason that the calculation procedure is the same for different visitors, the reasoning process and calculation details for Visitor 1 will be shown here only.

As shown in Table 2, the minimum and maximum values for the elements of ts, f and ti in their universal sets are:

$$\text{Minimum } ts = 32\,\text{s} \qquad \text{Maximum } ts = 196\,\text{s}$$
$$\text{Minimum } f = 1 \qquad \text{Maximum } f = 3$$
$$\text{Minimum } ti = 60\,\text{s} \qquad \text{Maximum } ti = 362\,\text{s}.$$

Table 1. Visiting time for various site pages.

Visitor 1 Tour	Visiting Time (s)	Visitor 2 Tour	Visiting Time (s)	Visitor 3 Tour	Visiting Time (s)
Com	96	SM	168	Off	180
Tele	32	F/W	100	F/W	200
F/W	50	Off	104	Tele	60
ID	20	Tele	180	Com	190
Off	60	Com	120	Off	60
ID	30	ID	60	F/W	200
BE	30	SM	180	SM	40
SM	80	BE	20	BE	60
Off	60	F/W	80	Com	140
Com	100				
BE	30				
Off	40				

Table 2. Data for various visitors associated with various favorite sites.

	Visitor 1			Visitor 2			Visitor 3		
Shopping Spots	Ts (s)	Fv	Ti (s)	Ts (s)	Fv	Ti (s)	Ts (s)	Fv	Ti (s)
Business Equipment	60	2	240	20	1	Inf.*	60	1	Inf.*
Computer	196	2	362	120	1	Inf.*	330	2	360
Food/Wine	50	1	Inf.*	180	2	664	400	2	310
Interior Design	50	2	60	60	1	Inf.*	0	0	Inf.*
Office	160	3	130/140	104	1	Inf.*	240	2	450
Shopping Mall	80	1	Inf.*	348	2	564	40	1	Inf.*
Telecommunication	32	1	Inf.*	180	1	Inf.*	60	1	Inf.*

*Inf. means infinity

In Table 2, the values of Ts, Fv and Ti for the three fuzzy sets are 60 s, 2 and 240 s respectively for the business equipment (BE) shop. The corresponding probabilities for the own set of Ts, Fv and Ti of the shop are listed in Table 3.

The output membership function W and the corresponding probability $\mu_W(w)$ can be determined by fuzzy rules (Eqs. (3) and (4a)) respectively. In this case, a total of twenty fuzzy rules are fired. However, only four rules can produce a non-zero probability value of the consequent, i.e. $\mu_Y > 0$. The resulting linguistic label of the four fuzzy rules and their corresponding value of μ_Y are given as:

IF $ts = \text{S}$ AND $f = \text{N}$ AND $ti = \text{RL}$ THEN $w = \#18$,
thus $\mu_{Y1} = 0.317 \bullet 1 \bullet 0.384 = 0.1217$

IF $ts = \text{S}$ AND $f = \text{N}$ AND $ti = \text{N}$ THEN $w = \#19$,
thus $\mu_{Y2} = 0.317 \bullet 1 \bullet 0.616 = 0.1953$

IF $ts = \text{RS}$ AND $f = \text{N}$ AND $ti = \text{RL}$ THEN $w = \#19$,
thus $\mu_{Y3} = 0.638 \bullet 1 \bullet 0.384 = 0.2623$

IF $ts = \text{RS}$ AND $f = \text{N}$ AND $ti = \text{N}$ THEN $w = \#20$,
thus $\mu_{Y4} = 0.683 \bullet 1 \bullet 0.616 = 0.4207$.

Procedure 3: Defuzzification

A crisp value can be obtained by finding the centroid of the polygon using Eq. (5). The determined crisp value is known as the ILIP for the BE shop for Visitor 1. Table 4 lists all numerical data and the result of the defuzzification process.

Table 3. The corresponding probabilities for the own set of Ts, Fv and Ti of the shop.

	Probability	Linguistic Label of Corresponding Membership Function
$ts = 60\,\text{s}$	$\mu_{Ts}(ts) = 0.683$	RS
	$\mu_{Ts}(ts) = 0.317$	S
$f = 2$	$\mu_{Fv}(f) = 1$	N
	$\mu_{Fv}(f) = 0$	RLO or RHI
$ti = 240\,\text{s}$	$\mu_{Ti}(ti) = 0.616$	N
	$\mu_{Ti}(ti) = 0.384$	RL

Table 4. Numerical data and the result of the defuzzification process.

Polygon	Area (A)	Moment Arm (r)	rA
1	0.1521	43.67	6.642
2	0.7241	43.99	31.853
3	1.0979	46.23	50.756
4	0.5259	48.33	25.417
	$\Sigma = 2.5$		$\Sigma = 114.668$

$$z = \frac{114.668}{2.5} = 45.8672$$

Repeat the procedures as stated in Secs. 4.2 and 4.3 so that the ILIP for the other shopping spots of Visitor 1 can be fully determined. Apply the same calculation method for Visitors 2 and 3, and a complete result for the three visitors are given in Table 5.

2.3.6. *Result analysis and discussion*

The calculated ILIP for each shopping spot of three visitors as listed in Table 5 can state the order of preferences of each visitor. For example, the order of shopping preferences for Visitors 1, 2 and 3 will be: office shop/interior design shop/computer shop/business equipment shop, shopping mall/food/wine shop, and food/wine shop/computer shop/office shop respectively. The ILIP values for each shopping spot can also indicate the relative level of interesting on the shops, such as Visitor 1 was in favor of shopping in the office shop much more than the shops of BE, ID and Com. The favorite shopping spot for Visitor 2 was shopping mall, however, both spots of F/W and Com were in favor by Visitor 3. These findings agree with the actual psychological decision of the visitors because the shopping spots with higher ILIP were visited again at the subsequent trips. Curiously, orders were also placed on these spots by tracing the records in the IIDS. Table 6 lists

Table 5. Complete result for the three visitors.

Shopping Spots	Visitor 1	Visitor 2	Visitor 3
Business Equipment	45.8672	—	—
Computer	50	—	94.4
Food/Wine	—	84.9	99.2
Interior Design	51.0992	—	—
Office	95.323	—	85.3
Shopping Mall	—	99.2	—
Telecommunication	—	—	—

Table 6. A set of touring and purchasing records of three visitors.

	Visitor 1				Visitor 2				Visitor 3			
Shopping Spots	*Ts* (min)	*Fv*	ILIP	No. of order	*Ts* (min)	*Fv*	ILIP	No. of order	*Ts* (min)	*Fv*	ILIP	No. of order
BE	522	15	85.5	2	210	10	86.1	1	23	3	7.5	0
Com	270	10	77.9	1	15	2	10.1	0	724	21	95.7	3
F/W	103	11	57.2	1	407	15	91.6	3	768	20	94.3	4
ID	440	19	87.7	2	78	6	37.4	0	110	4	19.2	0
Off	653	18	92.3	3	115	8	60.5	0	542	15	81.8	2
SM	221	8	56.1	0	388	13	89.7	3	209	12	44.4	1
Tele	78	5	22.3	0	147	7	60.2	1	441	17	76.8	2

a set of touring and purchasing records of three visitors within a month period of using the Cyber Mall Web site.

A few important findings were obtained from Tables 5 and 6 including:

(i) No ILIP can be determined for the spot which has a single visit in the same trip, because the visiting interval Ti becomes an infinity value as shown in Table 2.

(ii) For the spot with a longer total time of stay Ts and which is visited frequently in a trip, a bigger ILIP is always obtained.

(iii) The numerical value of ILIP can directly reflect the visitors' preference related to particular shopping spots. Promotion strategies are always put onto the spots with higher ILIP.

(iv) Switching of shopping favor among spots can always be detected by viewing the statistical data of ILIP over a period.

3. Conclusion

The rapid growth of Intranet/Internet technology has a significant impact on the operations of the manufacturing systems. In order to maintain the competitive edge in the manufacturing market, it is imperative that information can be accessed by employees, customers, suppliers and business partners electronically. This electronic information access lowers the cost of production internally and externally and it enables employees in your company to work more effectively together with customers, suppliers and business partners. However, before jumping onto the Intranet/Internet bandwagon, companies also need to consider the pros and cons of setting up such a system. Full support from the top management to the clerical staff is important if this technology is to be introduced.

References

1. N. V. Findler and G. D. Elder, Multi-agent coordination and cooperation in a distributed dynamic environment with limited resources, *Artificial Intelligence in Engineering* **9** (1995) 229–238.
2. H. Wang and C. Wang, Intelligent agents in the nuclear industry, *IEEE Computer* (1997) 28–34.
3. R. Bose, CMS: An intelligent knowledge-based tool for organizational procedure modeling and execution, *Expert Systems with Applications* **8** (1995) 1–21.
4. R. Bose, Intelligent agents framework for developing knowledge-based decision support systems for collaborative organizational processes, *Expert Systems with Applications* **11** (1996) 247–261.
5. P. E. Clements, R. M. Jones, R. H. Weston and E. A. Edmonds, A framework for the realization of cooperative systems, *SIGOIS Bulletin* **15** (1995) 9–10.
6. D.W. Ramsus, Intelligent agents: PC, *AI* **9** (1995) 27–32.
7. M. Stefik and D. G. Bobrow, Object-oriented programming: Themes and variations, *AI Magazine* **6** (1996) 40–62.
8. C. S. Krishnamoorthy and S. Rajeev, *Artificial Intelligence and Expert Systems for Engineers* (CRC Press, 1996).

9. CLIPS *Reference Manual*, Version 6.0 (1993).
10. J. Giarratano and G. Riley, *Expert Systems: Principles and Programming* (International Thompson Publishing, Boston, MA, 1993).
11. Microsoft Visual Basic, *Language Reference*, Microsoft Corporation (1997a).
12. Microsoft Visual Basic, *Programmer's Guide*, Microsoft Corporation (1997b).
13. Global Manufacturing Network, http://www.globalmfg.com, *Society of Manufacturing Engineers* (1998).
14. 3M Innovation Network, http://www.mmm.com, 3M, 1998.
15. Montgomery, Wong's Webshop, http://www.montywong.com (1998).
16. P. Ching, Push delivery: the direct approach, *Communique* **8** (1997) 1–12.
17. M. Rosen, Bring push technology to the masses, *Windows Sources*, Australia (1997).
18. A. Poler, *CyberAtlas, Push Technology*, http://cyberatlas.com/pushtec.html (1998).
19. R. Morochove, *Microsoft and Netscape Push Webcasting*, http://www.morochove.com/watch/cw/ff70320.htm (1998).
20. L. A. Zadeh, Fuzzy sets, *Information and Control* **8** (1965) 338–383.
21. A. Kaufman, Introduction to theory of fuzzy subsets (Academic, New York, 1975).
22. E. H. Mamdani, Applications of fuzzy algorithms for control of a simple dynamic plant, *Proc. IEEE* **121** (1974) 1585–1588.
23. M. Mizumoto, Note on the arithmetic rule by Zadeh for fuzzy reasoning methods, *Cyber System* **12** (1981) 247–306.
24. M. Mizumoto, Fuzzy controls by product-sum-gravity method, *Advancement of fuzzy theory and systems in China and Japan*, Beijing: IAP (1990).
25. J. Yan, M. Ryan and J. Power, *Using Fuzzy Logic* (Prentice Hall, New York, 1994).
26. Microsoft Site Server, 3.0., Microsoft Corporation. USA (1999).
27. E. Smith, Y. Malluf, J. McManus, A. Scott, C. Laird and M. C. Amundsen, *Inside VBScript with ActiveX*, New Riders (1997).

CHAPTER 7

AUTOMATED VISUAL INSPECTION: TECHNIQUES AND APPLICATIONS IN MANUFACTURING SYSTEMS

CHRISTOPHER C. YANG

Department of Systems Engineering and Engineering Management,
The Chinese University of Hong Kong,
Hong Kong
E-mail: yang@se.cuhk.edu.hk

Visual inspection plays an important role in computer aided and integrated manufacturing. A good inspection planning reduces the effort and improves the accuracy in inspection. In this chapter, we discuss the representation of three-dimensional parts using object-oriented representation and viewer-centered representation. Such representations support the planning of active visual inspection. In order to maximize the accuracy of active visual inspection, the quantization errors and displacement of active vision sensors are analyzed. The derivation of the probabilistic analysis is presented. Applications of such analysis are discussed.

Keywords: Active vision; inspection planning; aspect graph; quantization errors; displacement errors.

1. Introduction

Traditionally, the inspection of manufacturing products is done by using the coordinate measuring machines (CMM). Due to the advances in imaging technology, active vision, computers, and automated visual inspection are becoming more popular in industrial inspection. Circuit boards, electronic devices, and clothing are now inspected by vision systems. However, many visual inspection systems are still limited to two-dimensional objects. In order to achieve active vision inspection for three-dimensional parts, several inter-related problems must be solved. Representation of three-dimensional parts must be incorporated with the dimensional structure of the manufacturing features and the characteristics of the vision sensors. A good representation is important for inspection planning. In order to ensure the accuracy of inspection, the analysis of possible errors in visual inspection must also be investigated.

1.1. *Inspection planning*

Several approaches have been proposed for inspection planning based on CAD models. Menq *et al.*,[20,39] Merat and Radack,[22,23,26] ElMaraghy and Gu,[9] Chan and Gu,[5] have proposed inspection planning for coordinate measuring machines. Hutchinson *et al.*[14] and Park *et al.*[24,25] have proposed inspection planning for vision sensors. However, these works are limited to the use of coordinate measuring machines or passive vision. Visual inspection using active vision is not discussed in this paper.

Inspection planning[33,37,38] involves two major components, namely:

(i) Generating a list of measurable entities for an object based on the interaction between the features of the object;
(ii) Determining the camera location and orientation (sensor setting) for inspecting a set of entities.

In Sec. 2.1, the object-centered and viewer-centered representations of three-dimensional models are introduced. The geometric reasoning utilizes the object-centered representation to generate alternative inspection strategies based on the manufactured features and their spatial relationships to generate alternative inspection strategies. In generating the strategies, we determine a set of measurable entities, which, once they have been dimensioned, will allow us to provide a value for the length of the desired attribute and to decide upon a computation method based on the measurable entities. Measurable entities are topologic entities (edge segments) of the component's boundaries, whose true dimensions will be determined from the images of the object. These strategies are generated by rule matching procedures. The reasoning mechanism employed must be capable of generating alternative strategies to inspect the dimensional attributes of a component.

In Sec. 2.2, an entity-based aspect graph is introduced to determine potential sensor settings. An aspect graph is a graph-based representation of an object's characteristic views. An *entity-based aspect graph* provides the characteristic views for the measurable entities obtained from the inspection plan. Its complexity is lower than that of the traditional aspect graph. Construction of the entity-based aspect graph from scratch is discussed. Using the entity-based aspect graph approach, one can identify promising sensor settings for a given set of measurable entities by searching through the nodes in the graph. If the aspect graph methodology is not used, the boundaries of the visible regions are generated everytime when a set of measurable entities is given. The aspect graph methodology makes it more efficient to determine all the potential sensor settings.

1.2. *Error analysis*

Errors resulting from the displacement of the sensor and quantization in image digitization are inevitable in active vision. Several researchers have investigated these errors. Kamgar-Parsi,[16] Blostein,[1] Ho[13] and Griffin[12] have examined quantization errors. Su *et al.*,[29] Renders *et al.*,[27] Menq *et al.*,[19,21] Chen *et al.*,[6] Veitchegger *et al.*,[30] Bryson[4] and Smith *et al.*[28] have investigated the errors of robot manipulators. In Sec. 3, the analysis and modeling of the probability density functions of spatial quantization errors as well as the displacement errors are presented. The total errors in an active vision inspection are then calculated based on the integration of the quantization errors and the displacement errors. The expected accuracy in the inspection of a set of entities by a given sensor setting is then analyzed. Based on this analysis, one can determine whether a specific sensor setting is suitable for inspecting a specific dimension of a component in order to verify its design specifications.

2. Representation: Object-Oriented and Viewer-Centered

Inspecting a three-dimensional part requires knowledge of the three-dimensional structure of the manufacturing features and the characteristics of the vision sensors. An object-oriented representation, the Brep CAD model, and a viewer-centered representation, which is the entity-based aspect graph, are employed.

2.1. *Boundary representation (Brep)*

Boundary representation (Brep) is an object-oriented representation of a three-dimensional solid model. Brep provides an unambiguous representation of a solid's surface and its topological orientation. In Brep, the oriented surface of a solid is represented as a data structure composed of vertices, edges, and faces. Using the orientation of the solid surfaces, we can determine the solid's interior and exterior.

Brep consists of two descriptions, namely the topological description and the geometric description.[14] Topological description represents the connectivity and the orientation of vertices, edges, and faces. The geometric description represents for embedding the surface elements in space.

In topological representation, specific information must be provided for vertices, edges, and faces. The incident edges and adjacent faces have to be provided for each vertex. The adjacent faces form cones that can be nested. For each edge, the bounding vertices and its adjacent faces must be specified. The adjacent faces are ordered clockwise about the edge and are paired to enclose a wedge of solid interior. For each face, the bounding edges and vertices must be provided. The bounding edges and vertices are organized in a set of cycles enclosing the face area to the right.

The face direction vector is used for geometric representation. The face direction vector is a vector on the surface plane perpendicular to an edge tangent vector and pointing to the interior of the surface.

An object as shown in Fig. 1(b) depicts the Brep of the object in Fig. 1(a).

2.1.1. *Geometric reasoning for generating dimensioning strategies*

The Brep of an object provides the information regarding the semantic features and their spatial relationship of the object. Given this information, geometric reasoning determines the measurable entities to be dimensioned.

Using the topological and geometric information provided in Brep, geometric reasoning will first identify the abstract features of an object, such as through slot,

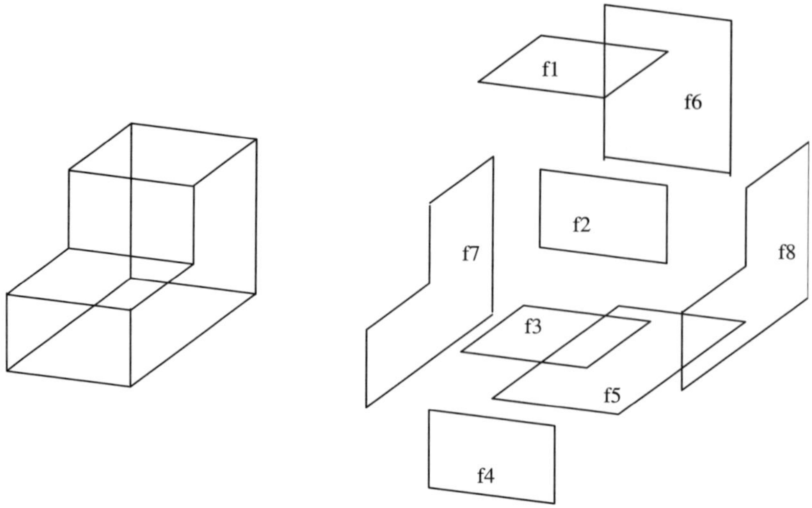

Fig. 1(a). An object with a step.

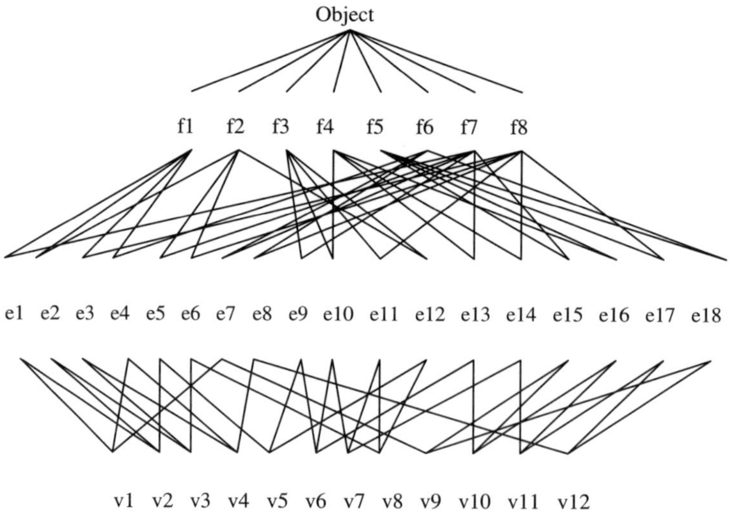

Fig. 1(b). Brep of the object in Fig. 1(a).

blind slot, through step, blind step and pocket, etc. Based on the identified features, the attributes that are necessary to be dimensioned will be captured. For example, if a through slot is identified, the length, width, and height of the slot should be dimensioned.

Given the attributes to be dimensioned for all the identified features of the object, geometric reasoning will then determine the strategies. A strategy is the computation methodology to dimension an attribute of a feature. Due to interaction

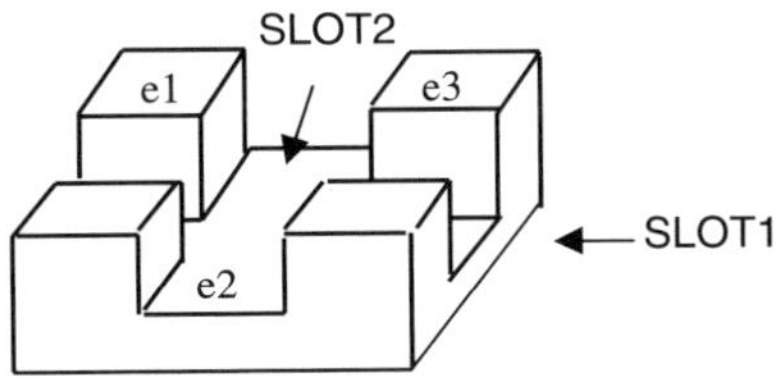

Fig. 2. An object with two features, SLOT1 and SLOT2.

between features, there may not be any existing edge that can be used to dimension an attribute. It may require a computation based on the dimensions of several existing edges. For example, in Fig. 2, the length of SLOT1, can be dimensioned by e1 + e2 + e3. There is probably more than one strategy for dimensioning an attribute. The geometric reasoning should be able to identify all possible strategies and find the most optimized strategy to dimension all the selected attributes.

2.2. *Aspect graph*

Aspect graph is a viewer-centered representation of an object based on concepts generally attributed to Koenderink and van Doorn.[17] Each node of an AG corresponds to a distinct characteristic view domain, and each arc between a pair of nodes represents the adjacency between the corresponding characteristic view domains. Each characteristic view domain has a "topologically distinct" view of the object. Algorithms have been derived to create AGs for three-dimensional objects by following specific rules to partition the three-dimensional space.

The problem with the AG is its complexity and its limited practical utility. As a result, there are limited applications of aspect graph on visual inspection. A simple object may have more than a hundred nodes in its aspect graph. The worst-case node complexity is $O\ (N^9)$ for an N-faced polyhedron. In Ref. 2, researchers have discussed why aspect graphs are not (yet) practical for computer vision in a panel. Therefore, an efficient viewer-centered representation of objects, which satisfies the requirements of its intended application, is desired.

Several modified aspect graphs have been developed to reduce the complexity and yet maintain the features that are necessary for its applications. For example, Eggert[10] has developed the scale space aspect graph and Laurentini[18] has developed the reduced aspect graph. The scale space aspect graph and the reduced aspect graph approaches reduce the size of the aspect graph. However, for the application of sensor placement in visual inspection and for object recognition, some unnecessary information is retained and some useful information is lost. Yang[32] has recently developed the entity-based aspect graph, which is specifically designed for the application of active vision inspection. In the following section, the general aspect graph and the entity-based aspect graphs will be introduced.

2.2.1. *Definition and construction of aspect graphs*

The AG is a double (C, A) where C is the set of the characteristic view domains represented as nodes, {C1, C2, C3, ...}, and A is the set of the adjacent pairs of characteristic view domains represented as arcs, {..., (C*i*, C*j*), ...}. Freeman[11] and Bowyer *et al.*[3] developed algorithms to construct AG for planar-faced solids. Basically, there are three types of partitioning rules for determining the characteristic view domains. Bowyer has described two visual events, the edge-vertex (EV) event and the edge-edge-edge (EEE) event. These visual events are similar to the partitioning types (type A, type B, type C) developed by Freeman. The EV events contains two categories. One of them involves edge-vertex pairs on the same face and the other involves pairs on separate faces. The first and second category of the EV event corresponds to the type A partition surfaces and type B partition barriers of Freeman's method, respectively. The EEE event is the general case where three edges share no common vertex and they correspond to the type C quadric surfaces. The partition rules introduced by Freeman and Bowyer divided the three-dimensional space into regions such that each region (characteristic view domain) has a distinct view of the object. Each partition plane or surface is constructed based on the edges and/or the vertices of the object to divide the space into two regions. From the vantage points of one region one is able to observe the corresponding edges or vertices but from the vantage points of the other region he is unable to do so.

Figure 3(b) shows an aspect graph for the object with a step feature as shown in Fig. 3(a).

2.2.2. *Entity-based aspect graph (EAG)*

The EAG is a quadruple (E, V, O, A), where E is a set of entities of interest for the object, {..., e*i*, ..., e*j*, ...}, such that the EAG only provides the observability of these entities. V is a set of viewing domains, {V1, V2, V3, ...}, in the three-dimensional space. O is the set of observable lists of entities for each element in V, $\{O_{V1}, O_{V2}, O_{V3}, \ldots\}$, where O_{Vi} is the list of observable entities of entity viewing domain V*i*. An entity is observable in an entity viewing domain only if no portion of this entity is occluded. Similar to the AG, A is the set of adjacent pairs of entity viewing domain, $\{\ldots, (Vi, Vj), \ldots\}$.

Given a set of entities of interest (EOI), one can use all the partition planes and surfaces of the object and eliminate those that are not required for constructing the EAG of the object. These partition planes or surfaces formed by the vertices and edges, which are not elements of the set of EOI, are eliminated. This is because the formed planes and surfaces partition only the regions having different observability with respect to the entities that are not in the EOI. The domain formed is possibly the combination of several characteristic-view domains of the AG. Although each domain may contain more than one characteristic view of the object, each of these characteristic views will observe the same entities in set E of the EAG.

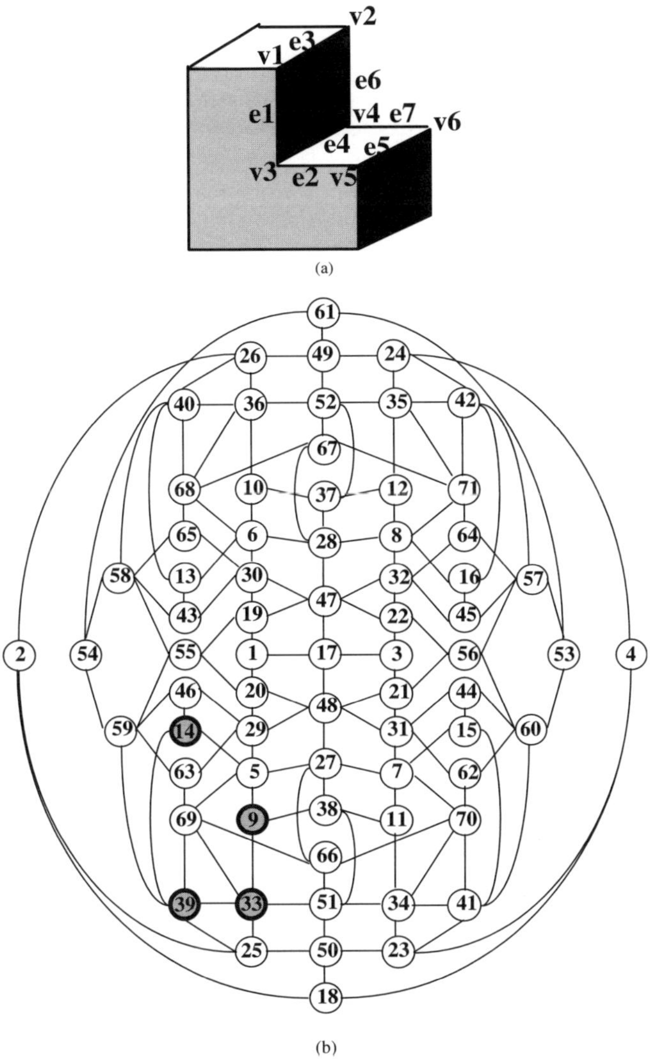

Fig. 3. (a) An object with a step, (b) aspect graph of object in (a).

Algorithm for constructing EAG with EOI = E

Input: The boundary representation of the object (vertices, edges, and faces).

Output: An EAG with a set of viewing domains, V, a set of lists of observable entities, O, a list for each element of V, and a set of adjacent pairs of viewing domain, A, i.e. EAG (E, V, O, A).

1. Find all the partition planes and surfaces (*using type A, type B and type C partition planes and surfaces as described by Freeman*[11]) and eliminate those generated by entities that are not elements of E, $\{p1, p2, \cdots, pn\}$.

2. Construct
   all possible n-tuples, L = $\{\{p1^+, p2^+, \ldots, pn^+\}, \{p1^-, p2^+, \ldots, pn^+\}, \ldots, \{p1^-, p2^-, \ldots, pn^-\}\}$ where pi^+, pi^- are inequalities denoting the different half spaces of the partition plane pi.
3. For each n-tuple,

   Determine the feasibility of the three-dimensional region described by the n-tuple. (*If a point cannot belong to the corresponding set of n inequalities, the region is infeasible.*)
   If such an n-tuple is feasible,

   Determine the exact boundaries of such a three-dimensional region. (*If all the intersections of an element in n-tuple (a half-space) with the other half spaces in the n-tuple are not on the boundary of the region represented by the n-tuple, then that element (half space) can be removed from the n-tuple.*)
   Save the reduced tuple in V.
   Save the corresponding list of observable entities in O. (*The list of observable entities is the union of entities observable from each* Pi^+ *in the n-tuple.*)
4. For each pair of elements in V,

   If these elements share the same boundary in the three-dimensional space,
   The corresponding regions are adjacent.
   Save the pair in A.

Figure 4 shows the EAG with E = $\{e1, e2, e3, e4, e5, e6, e7, v1, v2, v3, v4, v5, v6\}$ of the object in Fig. 3(a). The label of the entities is given in Fig. 3(a). The constructed AG has 71 nodes which is also the number reported by Freeman and Bowyer, while the EAG has only 20 nodes. To construct the EAG in Fig. 4 with the set of entities of interest labeled, 14 out of 16 possible partition planes are used. However, the number of nodes the EAG contains is only 28% of the number in the complete aspect graph. The complexity in the EAG is significantly reduced. For constructing the aspect graph for the step (Fig. 3(a)), we use eight type-A and eight type-B partition planes. The type-B partition planes are formed by eight edge-vertex pairs, namely: $\{e1, v6\}$, $\{e2, v2\}$, $\{e3, v5\}$, $\{e3, v6\}$, $\{e5, v1\}$, $\{e5, v2\}$, $\{e6, v5\}$ and $\{e7, v1\}$. Each of these edge-vertex pairs contains the elements of the entities of interest, therefore, none of them can be eliminated in constructing the EAG. The eight type-A partition planes actually correspond to the planes that are extended by the eight faces of the object. Six of these faces have some entities of interest along their boundaries, which means that the corresponding type-A partition planes are formed by the edge-vertex pairs, with the entities of interest as their elements. Therefore only the other two type-A partition planes can be eliminated in

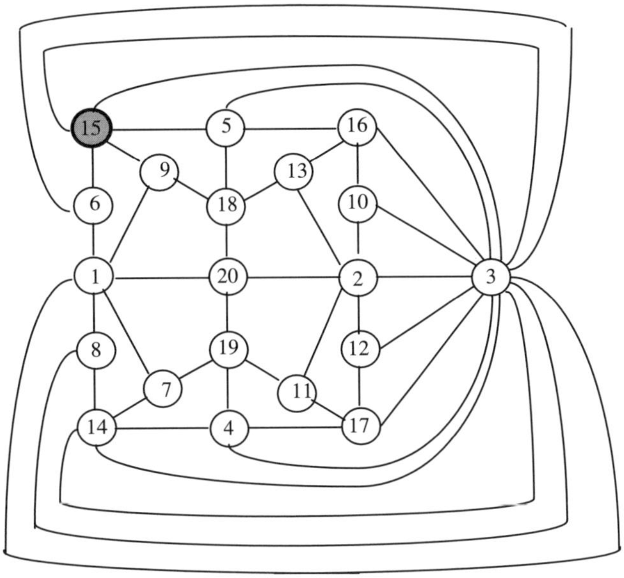

Fig. 4. Entity-based aspect graph of object in Fig. 3(a).

Table 1. The list of observable entities in each of the viewing domains of the entity-based aspect graph shown in Fig. 4.

Nodes	Visibility	Nodes	Visibility	Nodes	Visibility
O_1	e1, e2, v1, v3, v5	O_8	e1, e2, e3, v1, v2, v3, v5	O_{15}	e1, e2, e3, e5, v1, v2, v3, v5, v6
O_2	e6, e7v2, v4, v6	O_9	e1, e2, e3, v1, v2, v3, v5	O_{16}	e3, e5, e6, e7, v1, v2, v4, v5, v6
O_3	e1, e2, e3, e4, e5, e6, e7, v1, v2, v3, v4, v5, v6	O_{10}	e5, e6, e7, v2, v4, v5, v6	O_{17}	e3, e5, e6, e7, v1, v2, v4, v5, v6
O_4	e3, e5, v1, v2, v5, v6	O_{11}	e5, e6, e7, v2, v4, v5, v6	O_{18}	e3, v1, v2
O_5	e3, e5, v1, v2, v5, v6	O_{12}	e3, e6, e7, v1, v2, v4, v6	O_{19}	e5, v5, v6
O_6	e1, e2, e5, v1, v3, v5, v6	O_{13}	e3, e6, e7, v1, v2, v4, v6	O_{20}	None
O_7	e1, e2, e5, v1, v3, v5, v6	O_{14}	e1, e2, e3, e5, v1, v2, v3, v5, v6		

constructing the EAG. Table 1 presents the list of observable entities in the viewing domains for EAGs in Fig. 4.

Figure 5(a) shows an object with a pocket on the top face with all pocket edges labeled. If the entire pocket is a shape feature of interest, the 12 edges on the pocket are the elements of the set of EOI. In order to determine the possible characteristic views of these entities, an EAG with the set of entities, E = {L1, L2, L3, L4, W1, W2, W3, W4, H1, H2, H3, H4}, is constructed as shown in Fig. 5(b). The lists of observable entities in the viewing domains for EAGs in Fig. 5(b) is given in Table 2.

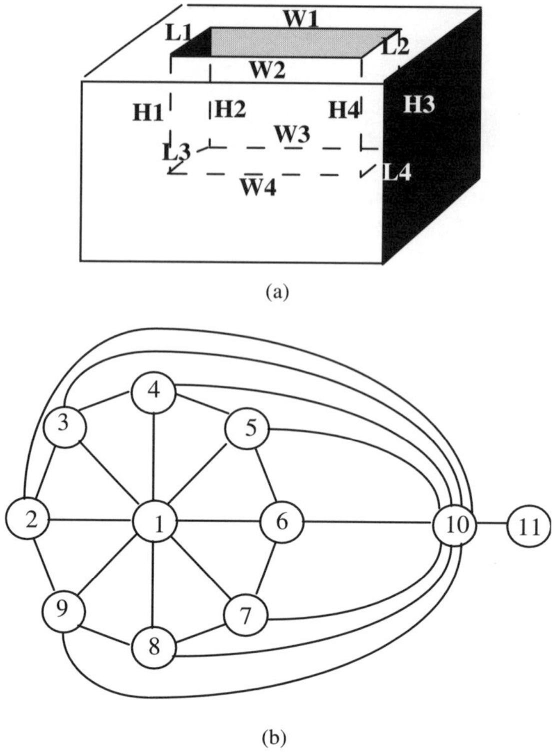

Fig. 5. (a) An object with a pocket on the top face. (b) The entity-based aspect graph of the object in Fig. 2(a) with the set of entities of interest as {L1, L2, L3, L4, W1, W2, W3, W4, H1, H2, H3, H4}.

Table 2. The sets of observable entities in each viewing domain of the entity-based aspect graph of the object in Fig. 2.

Node	Visibility	Node	Visibility
O_1	L1, L2, L3, L4, W1, W2, W3, W4, H1, H2, H3, H4	O_7	L1, L2, W1, W2, H2
O_2	L1, L2, L4, W1, W2, H3, H4	O_8	L1, L2, W1, W2, W3, H2, H3
O_3	L1, L2, W1, W2, H4	O_9	L1, L2, W1, W2, H3
O_4	L1, L2, W1, W2, W4, H1, H4	O_{10}	L1, L2, W1, W2
O_5	L1, L2, W1, W2, H1	O_{11}	None
O_6	L1, L2, L3, W1, W2, H1, H2		

2.2.3. *Application of entity-based aspect graph on active vision inspection*

In active vision inspection, we employ the entity-based aspect graph to determine the potential sensor settings for observing the topologic entities of three-dimensional components. A sensor setting determines the location and viewing direction where a noncontact sensor may be placed to observe one or more topologic entities whose dimensions are to be measured. The aspect graph of an inspected component provides all the possible characteristic views of the component. The entity-based aspect graph provides the characteristic views of the entities of interest on the component.

Geometric reasoning is used on the object as shown in Fig. 5(a). Here the identified feature is a pocket and the attributes that are necessary to be dimensioned are the height, length, and the width. The possible strategies for dimensioning the height of the step are {H1, H2, H3, or H4}. Similarly, the possible strategies for dimensioning the length and width of the step are {L1, L2, L3, or L4}, and {W1, W2, W3, or W4}, respectively. As a result, there are a total of 64 possible strategies for dimensioning the height, length, and width, i.e. {H1, L1, W1}, {H2, L1, W1}, {H3, L1, W1}, {H4, L1, W1}, etc.

Using the aspect graph representation, one may identify the possible locations of the vision sensor to dimension the measurable entities {H1, L1, W1} to be $\{O_1\}$, $\{O_4\}$, $\{O_6\}$, $\{O_2, O_4\}$, $\{O_2, O_5\}$, or $\{O_2, O_6\}$, etc. Although there are many potential sets of sensor locations, only some of them are able to dimension the measurable entities with an acceptable accuracy. For example, in O_1, the dimensions H1, L1, and W1 are all visible. However, the viewing direction is almost parallel to H1. A poor accuracy may be obtained if a vision sensor located in O_1 is used to dimension H1. As a result, we need to conduct an error analysis on all these potential strategies to ensure that the measurement accuracy reach the minimum requirement.

3. Error Analysis

In machine vision inspection, the geometric features measured are those that are unchanged with the environment or the set up of the vision system. Examples of such invariant features are the length, width, area and volume of a pocket. Measurements of invariant features by an automated, computer vision inspection system introduce errors in the observed dimensions. The sources of uncertainties that lead to these errors include the displacement of the camera (position and direction), the quantization error in image digitization, illumination errors (poor contrast and low intensity), the motion of the object or the camera setup, and parallax errors (object to camera distance too small).

Careful design and control of the environment can minimize to a negligible level the errors resulting from motion and parallax. However, the errors due to the displacement of the sensor, quantization errors in image digitization, and illumination errors cannot be avoided. They will always produce a significant effect on the measurement. In the following sections, the analysis of the quantization[35,36] and displacement[7,30] errors, as well as the integration[8,15,34] of such analysis with planning accuracy will be discussed.

3.1. *Planning accuracy and measurement accuracy*

Errors in the process of inspection will cause inaccuracies in the measurement of the inspected entities. Due to the displacement of the active vision sensor, the projections onto the image plane will not be at the expected location as computed from the given sensor setting. Due to the spatial quantization of the image, the dimensions of line segments will be quantized into a discrete number of pixels instead of the exact length projected onto the image plane. A careful analysis of the uncertainties that may exist in the inspection process can minimize the errors that will occur. We will categorize the analysis of inspection accuracy into two main aspects, *measurement* accuracy and *planning* accuracy.

Measurement accuracy involves the analysis of errors in the measurement of a manufactured product using a specific inspection strategy (for example, visual inspection, coordinate measurement machine, etc.). In this case, we analyze the accuracy using the measured value obtained in the inspection. For example, information from an image of a manufactured part is utilized in visual inspection. *Planning* accuracy involves the study of how the plan for inspection affects the accuracy of the inspection. In this case, we analyze the accuracy based on the inspection plan. The measured value of the product dimension(s) is not known because we have only the inspection plan without the execution of inspection. However, we do have the resolution of the camera and the planned sensor settings that can be used to inspect the dimensions for visual inspection. The error analysis of the dimensional measurement is based only on the probability density functions of the spatial quantization error of the image and the translational and orientational errors of the active vision head. This analysis gives us the ability to understand how to control the parameters of the sensor settings in order to increase the probability of high accuracy. In the following sections, we will study the *planning accuracy* problem in inspection.

3.2. *Quantization errors*

The spatial quantization error is important in inspection, especially when the size of the pixel is significant compared to the allowable tolerances in the object dimension. A quantized sample is indicated as a part of the object image if and only if more than half of it is covered by the edge segment. Significant distortion can be produced by this kind of quantization. A point on the image can only be located within one pixel of accuracy with traditional edge detection techniques.

Previous research has introduced some results on spatial quantization errors. Kamgar-Parsi[16] developed the mathematical tools for computing the average error due to quantization and he derived an analytic expression for the probability density of error of a function of an arbitrary large number of independent quantized variables. Blostein[1] analyzed the effect of image plane quantization on the 3D point position error obtained by triangulation from two quantized image planes in a stereo setup. Ho[13] expressed the digitizing error for various geometric features in terms of the dimensionless perimeter of the object and expanded the analysis to include the error in other measurements such as the orientation for both two-dimensional and three-dimensional measurements. Griffin[12] discussed an approach to integrate the errors inherent in the visual inspection such as spatial quantization errors, illumination errors and positional errors to determine the sensor capability for inspecting a specified part dimension using binary images. In the next sections, we shall investigate the quantization errors in visual inspection.

3.2.1. *One-dimensional spatial quantization errors*

In a one-dimensional environment, a straight line occupies a sequence of pixels with the width of each pixel represented by r_x. The range of the error in this case can be one pixel on either side of the line. Figure 6 shows an example where the number of pixels completely covered by the line is l. The length that partially covers the last pixel on each end of the line are u and v (where $u \leq r_x$ and $v \leq r_x$). The true length of the line, L, (before quantization) is $L = lr_x + u + v$.

We assume that u and v are independent and that each has a uniform distribution in the range $[0, r_x]$. The assumption of independence between u and v is valid because the actual length of the edge segment is not known, which means that the lengths partially covering u and v are uncertain. u and v are assumed to be uniformly distributed since the exact location of the terminal points are not known and the probability of the terminal point lying anywhere within a pixel is the same. Therefore their probability density functions (pdfs) are:

$$f_u(u) = \frac{1}{r_x} \quad \text{if } 0 \leq u \leq r_x;$$

and

$$f_v(v) = \frac{1}{r_x} \quad \text{if } 0 \leq v \leq r_x.$$

If $u \geq 1/2r_x$, the last pixel on the left will be quantized as part of the line. Otherwise, the pixel will not be considered as part of the line. Similar conditions hold for v on the right end of the line. The length of the line after quantization, L_q,

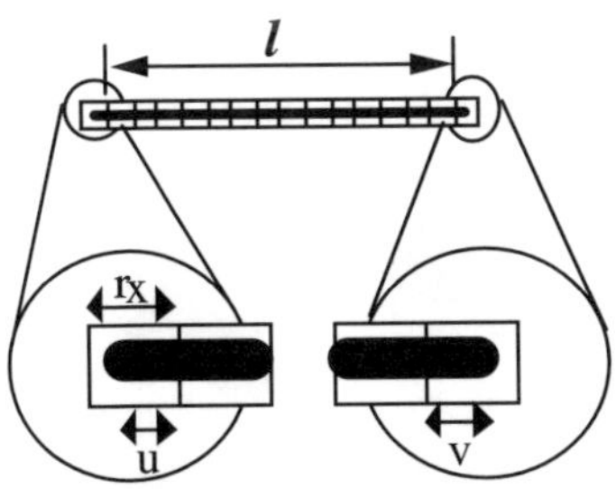

Fig. 6. A line of length $lr_x + u + v$ superimposed on a sequence of pixels of size r_x. The length of the line partially covering the last pixel on the left is u and the length of the line partially covering the last pixel on the right is v.

is given by:

$$L_q = \begin{cases} (l+2)r_x & \text{if } u \geq 0.5r_x \text{ and } v \geq 0.5r_x \\ (l+1)r_x & \text{if } (u \geq 0.5r_x \text{ and } v < 0.5r_x) \text{ or } (u < 0.5r_x \text{ and } v \geq 0.5r_x) \\ lr_x & \text{if } u < 0.5r_x \text{ and } v < 0.5r_x \end{cases}$$

The quantization error, ε_q, is defined as $L_q - L$. (Alternatively, ε_q can also be expressed as $\varepsilon_{q_u} + \varepsilon_{q_v}$ where ε_{q_u} is the quantization error of u and ε_{q_v} is the quantization error of v. ε_{q_u} and ε_{q_v} are each uniformly distributed with a range of $[-1/2r_x, +1/2r_x]$.) The expected value of ε_q is $\mathrm{E}[L_q] - \mathrm{E}[L]$. Since both u and v are uniformly distributed, the probability of each of the four conditions in the above equation is 1/4. Therefore, $\mathrm{E}[L_q]$ is $(l+1)r_x$. Since l and r_x are constants, both u and v have a uniform distribution between 0 and r_x. Their expected values are $1/2r_x$, and $\mathrm{E}[L]$ is $lr_x + r_x$. Hence, $\mathrm{E}[\varepsilon_q]$ is zero, and the range of the quantization error is $[-r_x, +r_x]$.

3.2.2. *Two-dimensional quantization errors*

The two-dimensional spatial quantization error is the combination of two one-dimensional spatial quantization errors. Figure 7 shows a line on a two-dimensional array of pixels. The resolution of the image is $r_x \times r_y$ where r_x is the width of a pixel and r_y is the length of a pixel. The horizontal component of the line length, L_x, is $l_x r_x + u_x + v_x$. The vertical component of the length of the line, L_y, is $l_y r_y + u_y + v_y$. The actual dimension, L, is therefore $\sqrt{L_x^2 + L_y^2}$.

The quantized length, $L_q = \sqrt{L_{q_x}^2 + L_{q_y}^2}$, where L_{q_x} and L_{q_y} are the horizontal and vertical quantized lengths, respectively. In one dimension, we have two random variables u and v, both of which have a uniform distribution. In two dimensions,

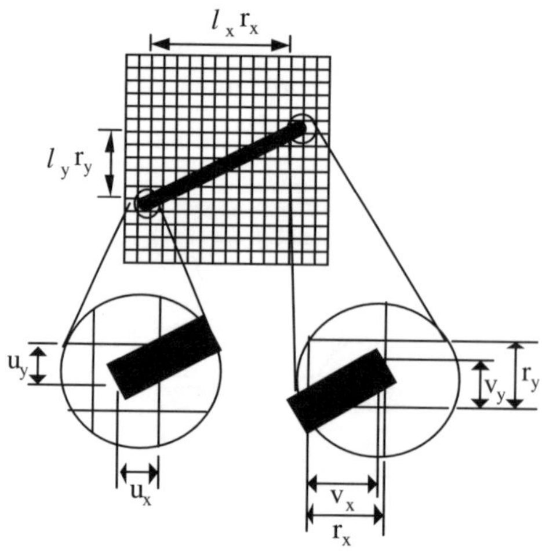

Fig. 7. A line on a two-dimensional array of pixels. The horizontal length of the line is $l_x r_x + u_x + v_x$. The vertical length of the line is $l_y r_y + u_y + v_y$.

there are four random variables, two for the horizontal length, u_x and v_x, and two for the vertical length, u_y and v_y. All four are assumed to be uniformly distributed.

$$f_{u_x}(u_x) = f_{v_x}(v_x) = \frac{1}{r_x} \quad \text{for } 0 \leq u_x \leq r_x \text{ or } 0 \leq v_x \leq r_x$$

$$f_{u_y}(u_y) = f_{v_y}(v_y) = \frac{1}{r_y} \quad \text{for } 0 \leq u_y \leq r_y \text{ or } 0 \leq v_y \leq r_y$$

A *geometric approximation* is used to characterize the two-dimensional quantization error. Figure 8 shows a line with length L lying at an angle of γ to the horizontal axis. (Note: The figure is not drawn to scale. ε_{q_x} is much smaller than L_x and ε_{q_y} is much smaller than L_y.) If $(\varepsilon_{q_x} + L_x)/(\varepsilon_{q_y} + L_y) = L_x/L_y$, the quantized line is parallel to the original line. In this case, the length of the quantized line, L_q, is $L + \varepsilon_{q_x} \cos\gamma + \varepsilon_{q_y} \sin\gamma$ as shown in Fig. 8. The two-dimensional spatial quantized error, $\varepsilon_q (= L_q - L)$, is $\varepsilon_{q_x} \cos\gamma + \varepsilon_{q_y} \sin\gamma$ in this case.

However, if the original line and the quantized line are not parallel, then $((\varepsilon_{q_x} + L_x)/(\varepsilon_{q_y} + L_y) \neq L_x/L_y)$, and the error, ε_q, is $\sqrt{(L\cos\gamma + \varepsilon_{q_x})^2 + (L\sin\gamma + \varepsilon_{q_y})^2}$. Although the lines may not be exactly parallel, they are approximately parallel because ε_x and ε_y are very small compared to the length of the line L. (The range of ε_x is $[-r_x, r_x]$ and the range of ε_y is $[-r_y, r_y]$ and the length of the line would typically be more than $100r_x$ or $100r_y$.) Therefore, $\varepsilon_q \cong \cos\gamma\ \varepsilon_{q_x} + \sin\gamma\ \varepsilon_{q_y}$.

Using this geometric approximation, we can compute the mean and the variance of the two-dimensional quantization error. The mean of the quantization error in two dimensions is:

$$\mathrm{E}[\varepsilon_q] \cong \mathrm{E}[\cos\gamma\ \varepsilon_{q_x} + \sin\gamma\ \varepsilon_{q_y}] = \cos\gamma\ \mathrm{E}[\varepsilon_{q_x}] + \sin\gamma\ \mathrm{E}[\varepsilon_{q_y}] = 0$$

since $\mathrm{E}[\varepsilon_{q_x}]$ and $\mathrm{E}[\varepsilon_{q_y}]$ are both zero. The variance of the quantization error in two dimensions is:

$$\mathrm{Var}[\varepsilon_q] \cong \mathrm{Var}[\cos\gamma\ \varepsilon_{q_x} + \sin\gamma\ \varepsilon_{q_y}] = \frac{1}{6}(\cos^2\gamma\ r_x^2 + \sin^2\gamma\ r_y^2)$$

since the variances of ε_{q_x} and ε_{q_y} are $1/6r_x^2$ and $1/6r_y^2$, respectively.

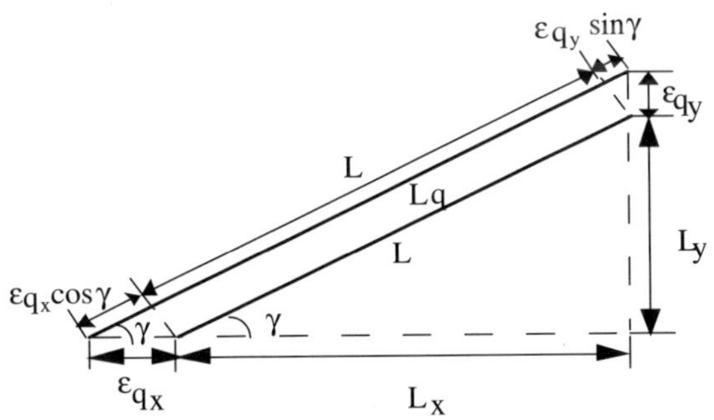

Fig. 8. The original line with length L. The angle between the line and the horizontal axis is γ. The quantized line has length L_q and it is parallel to the original line. (Figure is not shown in scale, ε_x and ε_y are much smaller than the length of the line in actual case.)

3.2.3. *Probability density function of quantization error*

The probability density function of the two-dimensional quantization error can be derived based on the geometric approximation. The pdf's of the errors in each dimension are as follows:

For the x-direction:

$$f_{\varepsilon_{qx}}(\varepsilon_{q_x}) = \begin{cases} -\frac{1}{r_x^2}\varepsilon_{q_x} + \frac{1}{r_x} & 0 \le \varepsilon_{q_x} \le r_x \\ \frac{1}{r_x^2}\varepsilon_{q_x} + \frac{1}{r_x} & -r_x \le \varepsilon_{q_x} \le 0 \\ 0 & \text{else} \end{cases} .$$

For the y-direction:

$$f_{\varepsilon_{qy}}(\varepsilon_{q_y}) = \begin{cases} -\frac{1}{r_y^2}\varepsilon_{q_y} + \frac{1}{r_y} & 0 \le \varepsilon_{q_y} \le r_y \\ \frac{1}{r_y^2}\varepsilon_{q_y} + \frac{1}{r_y} & -r_y \le \varepsilon_{q_y} \le 0 \\ 0 & \text{else} \end{cases} .$$

The approximate two-dimensional spatial quantization error, $\tilde{\varepsilon}$, is in terms of two variables ε_{q_x} and ε_{q_y}. The pdf is expressed in terms of $\tilde{\varepsilon}$ and the joint statistics of ε_{q_x} and ε_{q_y}.

We define z as:

$$z = \varepsilon_{q_x} + \varepsilon_{q_y}$$

and

$$\hat{z} = \varepsilon_q,$$

then,

$$\varepsilon_{q_x} = \frac{\tilde{\varepsilon}_q - \sin\gamma\ \hat{z}}{\cos\gamma}.$$

and

$$f_{\tilde{\varepsilon}_q,\hat{z}}(\tilde{\varepsilon}_q,\hat{z}) = J(\varepsilon_{q_x},\varepsilon_{q_y})^{-1} f\left(\frac{\tilde{\varepsilon}_q - \sin\gamma\ \varepsilon_{q_y}}{\cos\gamma}, \varepsilon_{q_y}\right)$$

where,

$$J(\varepsilon_{q_x},\varepsilon_{q_y}) = \begin{vmatrix} \frac{\partial\tilde{\varepsilon}_q}{\partial\varepsilon_{q_x}} & \frac{\partial\tilde{\varepsilon}_q}{\partial\varepsilon_{q_y}} \\ \frac{\partial\hat{z}}{\partial\varepsilon_{q_x}} & \frac{\partial\hat{z}}{\partial\varepsilon_{q_y}} \end{vmatrix} = |\cos\gamma|.$$

Therefore,

$$f_{\tilde{\varepsilon}_q}(\tilde{\varepsilon}_q) = \frac{1}{|\cos\gamma|}\int_{-\infty}^{\infty} f\left(\frac{\tilde{\varepsilon}_q - \sin\gamma\ \varepsilon_{q_y}}{\cos\gamma}, \varepsilon_{q_y}\right) d\varepsilon_{q_y}.$$

If ε_{q_x} and ε_{q_y} are independent, then $f(\varepsilon_{q_x},\varepsilon_{q_y}) = f_{\varepsilon_{q_x}}(\varepsilon_{q_x}) f_{\varepsilon_{q_y}}(\varepsilon_{q_y})$. Substituting this into the above equation, we obtain the probability density function of the spatial quantization error:

$$f_{\tilde{\varepsilon}_q}(\tilde{\varepsilon}_q) = \frac{1}{|\cos\gamma|}\int_{-\infty}^{\infty} f_{\varepsilon_{qx}}\left(\tan\gamma\left(\frac{1}{\sin\gamma}\tilde{\varepsilon}_q - \tau\right)\right) f_{\varepsilon_{qy}}(\tau)\,d\tau.$$

3.3. *Displacement errors*

In addition to quantization errors in active vision inspection, the uncertainties arising from robot motion and sensory information is also important. Su *et al.*[29] presented a methodology for manipulating and propagating spatial uncertainties in a robotic assembly system for generating executable actions for accomplishing a desired task. Renders *et al.*[27] presented a robot calibration technique based on a maximum likelihood approach for the identification of errors in position and orientation. Menq *et al.*[21] presented a framework to characterize the distribution of the position errors of robot manipulators over the work space and to study their statistical properties. Chen *et al.*[6] identified and parameterized the sources that contribute to the positioning error and they estimated the values of the parameters. Veitschegger *et al.*[30] presented a calibration algorithm for finding the values of the independent kinematic errors by measuring the end-effector Cartesian positions. In addition they have proposed two compensation algorithms: a differential error transform compensation algorithm and a Newton–Raphson compensation algorithm. Bryson[4] discussed methods for the measurement and characterization of the static distortion of the position data from 3D trackers including least-squares polynomial fit calibration, linear lookup calibration, and bump lookup calibration. Smith *et al.*[28] presented a method for explicitly representing and manipulating the uncertainty associated with the transformation between coordinate frames that represent the relative locations of objects. Sensors are used to reduce this uncertainty.

In active vision inspection, different sensor settings are used to position the active head to obtain and inspect various dimensions of interest. In placing the active head to achieve this task, errors in the final position and orientation are common. An end-effector has six degrees of freedom: three translations and three orientations. The image of the inspected part obtained from the sensor, the entities observable from that image, and the measured dimensions of the entities are all dependent on the sensor location and viewing direction. If the sensor location and orientation are different from the planned sensor setting (i.e. there is *sensor displacement*), the same entities may be observable, but as a result of the displacement, the dimensions derived from the image will be inaccurate. The difference between the observed dimensions and the actual dimensions is defined as *displacement error*. The analysis in this section yields a better understanding of dimensional measurement incorporating error due to the displacement of the active sensor. These results should be useful in minimizing the occurrence and impact of such errors.

3.3.1. *Translational and orientational errors in perspective images*

In active vision inspection, the desired positions and orientations of the head are fed into the server control which accomplishes the sensor placement task by a sequence of movements. The horizontal and vertical displacements of the projected points on the two-dimensional image are affected by the six orientation and translation parameters of the sensor setting, the focal length of the sensor, the translational and orientational errors, and the three-dimensional coordinates of the model.

Given the orientation and location of the sensor and the coordinates of the object, we can compute the projected image of the object as follows:

$$\begin{bmatrix} x_I \\ y_I \\ z_I \\ c \end{bmatrix} = P_{\text{per}} Q \begin{bmatrix} x_w \\ y_w \\ z_w \\ 1 \end{bmatrix},$$

$$u = x_I / c,$$

and

$$v = y_I / c,$$

where $[x_w, y_w, z_w]^{\mathrm{T}}$ are the coordinates of the object in the world coordinate system, $(u,\ v)$ are the projected coordinates in the image plane, P_{per} is the matrix for perspective projection, and Q is the transformation matrix between the world coordinates and the image coordinate system.

$$P_{\text{per}} = \begin{bmatrix} 1 & 0 & 0 & 0 \\ 0 & 1 & 0 & 0 \\ 0 & 0 & 1 & -\frac{1}{f} \\ 0 & 0 & 0 & 1 \end{bmatrix} \quad \text{and} \quad Q = \begin{bmatrix} r_{11} & r_{12} & r_{13} & t_x \\ r_{21} & r_{22} & r_{23} & t_y \\ r_{31} & r_{32} & r_{33} & t_z \\ 0 & 0 & 0 & 0 \end{bmatrix},$$

where f is the focal length of the sensor, $[r_{ij}]$ 3×3 is the rotation submatrix in terms of the orientation parameters and $[t_k]$ 3×1 is the translation submatrix in terms of the translation parameters.

If the sensor is displaced, the correct coordinates of the projected points must be computed with modifications of the Q matrix because the rotation and translation parameters are distorted due to displacement. Thus, a matrix Q' must be substituted for Q to compensate for this. Q' is expressed in terms of the three translational errors, dx, dy, dz, and the three orientational errors, δx, δy, and δz, as well as the original translational and orientational parameters. The perspective matrix is unchanged since its only parameter, f, is fixed.

$$Q' = Q + \Delta Q = \text{Trans}(dx, dy, dz)\ \text{Rot}(x, \delta x)\ \text{Rot}(y, \delta y)\ \text{Rot}(z, \delta z) Q,$$

where $\text{Trans}(dx, dy, dz)$ is a transformation representing a translation by dx, dy and dz. $\text{Rot}(x, \delta x)$, $\text{Rot}(y, \delta y)$, and $\text{Rot}(z, \delta z)$ are transformations representing a differential rotation about the x, y and z directions respectively. Δ is given as:

$$\Delta = \begin{bmatrix} 0 & -\delta z & \delta y & dx \\ \delta z & 0 & -\delta x & dy \\ -\delta y & \delta x & 0 & dz \\ 0 & 0 & 0 & 0 \end{bmatrix}.$$

Therefore,

$$Q' = Q + \Delta Q$$
$$= \begin{bmatrix} r_{11} - r_{21}\delta z + r_{31}\delta y & r_{12} - r_{22}\delta z + r_{32}\delta y & r_{13} - r_{23}\delta z + r_{33}\delta y & t_x - \delta z t_y + \delta y t_z + dx \\ r_{21} + r_{11}\delta z - r_{31}\delta x & r_{22} + r_{12}\delta z - r_{32}\delta x & r_{23} + r_{13}\delta z - r_{33}\delta x & t_y + \delta z t_x - \delta x t_z + dy \\ r_{31} - r_{11}\delta y + r_{21}\delta x & r_{32} - r_{12}\delta y + r_{22}\delta x & r_{33} - r_{13}\delta y + r_{23}\delta x & t_z - \delta y t_x + \delta x t_y + dz \\ 0 & 0 & 0 & 1 \end{bmatrix}.$$

As a result, the displaced image coordinates are (u', v') are:

$$u' = \frac{f(\mathrm{C}1 + \mathrm{C}3\delta y - \mathrm{C}2\delta z + dx)}{f - (\mathrm{C}3 + \mathrm{C}2\delta x - \mathrm{C}1\delta y + dz)},$$

and

$$v' = \frac{f(\mathrm{C}2 + \mathrm{C}1\delta z - \mathrm{C}3\delta x + dy)}{f - (\mathrm{C}3 + \mathrm{C}2\delta x - \mathrm{C}1\delta y + dz)},$$

where

$$\mathrm{C}1 = r_{11}x_w + r_{12}y_w + r_{13}z_w + t_x;$$
$$\mathrm{C}2 = r_{21}x_w + r_{22}y_w + r_{23}z_w + t_y;$$

and

$$\mathrm{C}3 = r_{31}x_w + r_{32}y_w + r_{33}z_w + t_z.$$

The image coordinates $(u,\, v)$ without the displacement of the head are:

$$u = \frac{f\mathrm{C}1}{f - \mathrm{C}3},$$

and

$$v = \frac{f\mathrm{C}2}{f - \mathrm{C}3}.$$

The horizontal and vertical displacement errors, ε_{d_u} and ε_{d_v}, are as follows:

$$\varepsilon_{d_u} = u' - u = \frac{[\lambda_1\ \lambda_2\ \lambda_3 \lambda_4\ \lambda_5\ \lambda_6\ \lambda_7][\delta x\ \delta y\ \delta z\ dx\ dy\ dz]^{\mathrm{T}}}{[\lambda_{15}\ \lambda_{16}\ \lambda_{17}\ \lambda_{18}\ \lambda_{19}\ \lambda_{20}\ \lambda_{21}][\delta x\ \delta y\ \delta z\ dx\ dy\ dz]^{\mathrm{T}}},$$
$$\varepsilon_{d_v} = v' - v = \frac{[\lambda_8\ \lambda_9\ \lambda_{10}\ \lambda_{11}\ \lambda_{12}\ \lambda_{13}\ \lambda_{14}][\delta x\ \delta y\ \delta z\ dx\ dy\ dz]^{\mathrm{T}}}{[\lambda_{15}\ \lambda_{16}\ \lambda_{17}\ \lambda_{18}\ \lambda_{19}\ \lambda_{20}\ \lambda_{21}][\delta x\ \delta y\ \delta z\ dx\ dy\ dz]^{\mathrm{T}}},$$

where

$$\begin{array}{lll}
\lambda_1 = f\text{C1C2} & \lambda_2 = f((f-\text{C3})\text{C3}-\text{C1}^2) & \lambda_3 = f\text{C2}(\text{C3}-f) \\
\lambda_4 = f(f-\text{C3}) & \lambda_5 = 0 & \lambda_6 = f\text{C1} \\
\lambda_7 = 0 & \lambda_8 = f(\text{C2}^{2-}(f-\text{C3})\text{C3}) & \lambda_9 = -f\text{C1C2} \\
\lambda_{10} = f\text{C1}(f-\text{C3}) & \lambda_{11} = 0 & \lambda_{12} = f(f-\text{C3}) \\
\lambda_{13} = f\text{C2} & \lambda_{14} = 0 & \lambda_{15} = -\text{C2}(f-\text{C3}) \\
\lambda_{16} = \text{C1}(f-\text{C3}) & \lambda_{17} = 0 & \lambda_{18} = 0 \\
\lambda_{19} = 0 & \lambda_{20} = f-\text{C3} & \lambda_{21} = (f-\text{C3}).
\end{array}$$

The displacement errors are expressed in terms of the focal length of the sensor, the three-dimensional world coordinates, (x_w, y_w, z_w), the translation and orientation parameters of the sensor, and the translational and orientational errors of the active head $[dx, dy, dz, \delta x, \delta y, \delta z]$. As a result, for different three-dimensional points projected onto an image plane, the horizontal and vertical displacement errors on the image are not the same even if the focal length of the sensor, as well as the translational and orientational errors of the manipulator remain the same. For instance, two points on a three-dimensional object, (x_1, y_1, z_1) and (x_2, y_2, z_2), may have unequal horizontal displacement errors, $\varepsilon_{d_{u1}}$ and $\varepsilon_{d_{u2}}$, and unequal vertical displacement errors, $\varepsilon_{d_{v1}}$ and $\varepsilon_{d_{v2}}$. This implies that the distribution of the displacement error for each projected point in an image is unique in spite being generated by the same distribution of translational and orientational errors of the sensor.

3.3.2. *Probability density function of displacement errors*

Suppose the uncertainties in translation and orientation errors are all normally distributed with zero mean such that

$$f_a(a) = \frac{1}{\sqrt{2\pi}\sigma_a} e^{-a^2/(2\sigma a^2)},$$

where a represents any one of the translation errors dx, dy, dz, or the orientation errors δx, δy, δz. Normal distributions allow simple propagation of the errors with linear relationship. The normal distribution is also commonly used in error propagation for robot manipulators. Since the image coordinate errors ε_{d_u} and ε_{d_v} are expressed in terms of the six translation and orientation errors, we compute the probability density function of these errors as follows:

If $a = \beta_1 a_1 + \beta_2 a_2 + \cdots + \beta_n a_n + \alpha$ where a_i are all normally distributed with zero mean and independent for $i = 1, 2, \ldots, n$ and α is a constant, then

$$f_a(a) = \frac{1}{\sqrt{2\pi(\beta_1^2\sigma_{a_1}^2 + \beta_2^2\sigma_{a_2}^2 + \cdots + \beta_n^2\sigma_{a_n}^2)}} e^{-(a-\alpha)^2/(2(\beta_1^2\sigma_{a_1}^2+\beta_2^2\sigma_{a_2}^2+\cdots+\beta_n^2\sigma_{a_n}^2))},$$

where $\mu_a = \alpha$, and $\alpha_a^2 = \beta_1^2\sigma_{a_1}^2 + \beta_2^2\sigma_{a_2}^2 + \cdots + \beta_n^2\sigma_{a_n}^2$.

The horizontal and vertical displacement errors, ε_{d_u} and ε_{d_v}, are rational functions where both the numerator and the denominator are expressed in terms of three rotational errors and three translational errors of the active head, as well as a constant.

Let

$$\varepsilon_{d_u} = \frac{\zeta}{\chi},$$

and

$$\varepsilon_{d_v} = \frac{\xi}{\chi},$$

where

$$\begin{aligned}\zeta &= [\lambda_1\ \lambda_2\ \lambda_3\ \lambda_4\ \lambda_5\ \lambda_6\ \lambda_7][\delta x\ \delta y\ \delta z\ dx\ dy\ dz],\\ \xi &= [\lambda_8\ \lambda_9\ \lambda_{10}\ \lambda_{11}\ \lambda_{12}\ \lambda_{13}\ \lambda_{14}][\delta x\ \delta y\ \delta z\ dx\ dy\ dz],\\ \chi &= [\lambda_{15}\ \lambda_{16}\ \lambda_{17}\ \lambda_{18}\ \lambda_{19}\ \lambda_{20}\ \lambda_{21}][\delta x\ \delta y\ \delta z\ dx\ dy\ dz].\end{aligned}$$

The probability density functions of the numerator and the denominator can be derived as follows:

$$f_\zeta(\zeta) = \frac{1}{\sqrt{2\pi}\sigma_\zeta} e^{-(\zeta-\mu_\zeta)^2/(2\sigma_\zeta^2)},$$

where $\mu_\zeta = \lambda = 0$ and $\sigma_\zeta^2 = \lambda_1^2\sigma_{\delta x}^2 + \lambda_2^2\sigma_{\delta y}^2 + \lambda_3^2\sigma_{\delta z}^2 + \lambda_4^2\sigma_{dx}^2 + \lambda_5^2\sigma_{dy}^2 + \lambda_6^2\sigma_{dz}^2$;

$$f_\xi(\xi) = \frac{1}{\sqrt{2\pi}\sigma_\xi} e^{-(\xi-\mu_\xi)^2/(2\sigma_\xi^2)},$$

where $\mu_\xi = \lambda_{14} = 0$ and $\sigma_\xi^2 = \lambda_8^2\sigma_{\delta x}^2 + \lambda_9^2\sigma_{\delta y}^2 + \lambda_{10}^2\sigma_{\delta z}^2 + \lambda_{11}^2\sigma_{dx}^2 + \lambda_{12}^2\sigma_{dy}^2 + \lambda_{13}^2\sigma_{dz}^2$;

$$f_\chi(\chi) = \frac{1}{\sqrt{2\pi}\sigma_\chi} e^{-(\chi-\mu_\chi)^2/(2\sigma_\chi^2)},$$

where $\mu_\chi = \lambda_{21} = (f - \mathrm{C3})^2$ and $\sigma_\chi^2 = \lambda_{15}^2\sigma_{\delta x}^2 + \lambda_{16}^2\sigma_{\delta y}^2 + \lambda_{17}^2\sigma_{\delta z}^2 + \lambda_{18}^2\sigma_{dx}^2 + \lambda_{19}^2\sigma_{dy}^2 + \lambda_{20}^2\sigma_{dz}^2$.

Since ζ, ξ, and χ are all expressed in terms of the same translational and rotational errors, they are dependent on one another. To find the probability density function of ε_{d_u} and ε_{d_v}, the correlation coefficients of ζ and χ, as well as ξ and χ are needed.

The correlation coefficient of of ζ and χ is:

$$r_{\xi,\chi} = \frac{\mathrm{Cov}(\zeta,\chi)}{\sigma_\zeta\sigma_\chi},$$

where $\mathrm{Cov}(\zeta,\chi) = \lambda_1\lambda_{15}\sigma_{\delta x}^2 + \lambda_2\lambda_{16}\sigma_{\delta y}^2 + \lambda_3\lambda_{17}\sigma_{\delta z}^2 + \lambda_4\lambda_{18}\sigma_{dx}^2 + \lambda_5\lambda_{19}\sigma_{dy}^2 + \lambda_6\lambda_{20}\sigma_{dz}^2$.

The correlation coefficient of ξ and χ is:

$$r_{\xi,\chi} = \frac{\mathrm{Cov}(\xi,\chi)}{\sigma_\xi\sigma_\chi},$$

where $\mathrm{Cov}(\xi,\chi) = \lambda_8\lambda_{15}\sigma_{\delta x}^2 + \lambda_9\lambda_{16}\sigma_{\delta y}^2 + \lambda_{10}\lambda_{17}\sigma_{\delta z}^2 + \lambda_{11}\lambda_{18}\sigma_{dx}^2 + \lambda_{12}\lambda_{19}\sigma_{dy}^2 + \lambda_{13}\lambda_{20}\sigma_{dz}^2$.

Let

$$g_1(\varepsilon_{d_u}) = \sigma_\zeta^2 - 2r_{\zeta,\chi}\sigma_\zeta\sigma_\chi\varepsilon_{d_u} + \sigma_\chi^2\varepsilon_{d_u}, \quad g_2(\varepsilon_{d_u}) = \sigma_\zeta - r_{\zeta,\chi}\sigma_\chi\varepsilon_{d_u},$$
$$h_1(\varepsilon_{d_v}) = \sigma_\xi^2 - 2r_{\xi,\chi}\sigma_\xi\sigma_\chi\varepsilon_{d_v} + \sigma_\chi^2\varepsilon_{d_v}, \text{ and } h_2(\varepsilon_{d_v}) = \sigma_\xi - r_{\xi,\chi}\sigma_\chi\varepsilon_{d_v}.$$

Then,

$$f_{\varepsilon_{d_u}}(\varepsilon_{d_u}) = \frac{\sigma_\zeta\sigma_\chi\sqrt{1-r_{\zeta,\chi}^2}}{\pi g_1(\varepsilon_{d_u})}\exp\left(-\frac{\mu_\chi^2}{2g_1(\varepsilon_{d_u})}\left(\varepsilon_u^2 + \frac{g_2(\varepsilon_{d_u})^2}{\sigma_\chi^2}\right)\right)$$
$$+ \frac{g_2(\varepsilon_{d_u})\mu_\chi\sigma_\zeta}{\sqrt{2\pi}g_1(\varepsilon_{d_u})^{3/2}}\exp\left(-\frac{\mu_\chi^2\varepsilon_{d_u}^2}{2g_1(\varepsilon_{d_u})}\right)\operatorname{erf}\left(\frac{g_2(\varepsilon_{d_u})\mu_\chi}{\sigma_\chi\sqrt{2(1-r_{\zeta,\chi}^2)g_1(\varepsilon_{d_u})}}\right),$$

and

$$f_{\varepsilon_{d_v}}(\varepsilon_{d_v}) = \frac{\sigma_\xi\sigma_\chi\sqrt{1-r_{\xi,\chi}^2}}{\pi h_1(\varepsilon_{d_v})}\exp\left(-\frac{\mu_\chi^2}{2h_1(\varepsilon_{d_v})}\left(\varepsilon_{d_v}^2 + \frac{h_2(\varepsilon_{d_v})^2}{\sigma_\chi^2}\right)\right)$$
$$+ \frac{h_2(\varepsilon_{d_v})\mu_\chi\sigma_\xi}{\sqrt{2\pi}h_1(\varepsilon_{d_v})^{3/2}}\exp\left(-\frac{\mu_\chi^2\varepsilon_{d_v}^2}{2h_1(\varepsilon_{d_v})}\right)\operatorname{erf}\left(\frac{h_2(\varepsilon_{d_v})\mu_\chi}{\sigma_\chi\sqrt{2(1-r_{\xi,\chi}^2)g_1(\varepsilon_{d_v})}}\right).$$

3.3.3. *Displacement errors in dimensional measurement*

The displacement errors in projected points can result in errors in the measured area of a surface, the measured curvature of an arc or the measured length of a line segment, since these features are composed of the projection of the corresponding points from the three-dimensional model. The analysis of the displacement errors in these measurements is very important in visual inspection and other computer vision tasks such as gazing, pattern recognition etc. In this section, we will investigate the displacement errors introduced in the dimension measurement of linear segments.

The coordinates of the end-points of a linear segment in an image are ($u1$, $v1$) and ($u2$, $v2$). The length of the line segment is the distance between the end-points. Due to the displacement of the sensor, the end-points are not projected onto the desired locations ($u1$, $v1$) and ($u2$, $v2$). Instead, they are displaced to ($u1'$, $v1'$) and ($u2'$, $v2'$). Different translational and orientational errors of the sensor placement result in different displacements of the projected segment as illustrated in Fig. 9. For fixed translational and orientational errors of the sensor, the values of the horizontal and vertical displacement errors of $u1$ and $u2$ are not identical. Similarly the displacement errors of $v1$ and $v2$ are also not identical.

Let ($u1$, $v1$) correspond to the projection of model point ($x1$, $y1$, $z1$) and ($u2$, $v2$) correspond to the projection of model point ($x2$, $y2$, $z2$). The horizontal and

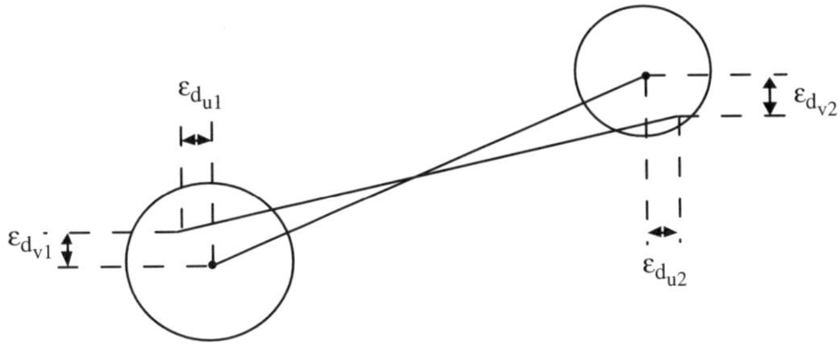

Fig. 9. The horizontal errors, $\varepsilon_{d_{u1}}$ and $\varepsilon_{d_{u2}}$ and the vertical errors, $\varepsilon_{d_{v1}}$ and $\varepsilon_{d_{v2}}$, due to the displacement of the camera.

vertical displacement errors of ($u1$, $v1$) and ($u2$, $v2$) can be computed from:

$$\varepsilon_{d_{u_i}} = \frac{[\lambda_1^i\ \lambda_2^i\ \lambda_3^i\ \lambda_4^i\ 0\ \lambda_6^i][\delta x\ \delta y\ \delta_z\ dx\ dy\ dz]^{\mathrm{T}}}{[\lambda_{15}^i\ \lambda_{16}^i\ 0\ 0\ 0\ \lambda_{20}^i\ \lambda_{21}^i][\delta x\ \delta y\ \delta z\ dx\ dy\ dz]^{\mathrm{T}}}$$

$$\varepsilon_{d_{v_i}} = \frac{[\lambda_8^i\ \lambda_9^i\ \lambda_{10}^i\ 0\ \lambda_{12}^i\ \lambda_{13}^i][\delta x\ \delta y\ \delta z\ dx\ dy\ dz]^{\mathrm{T}}}{[\lambda_{15}^i\ \lambda_{16}^i\ 0\ 0\ 0\ \lambda_{20}^i\ \lambda_{21}^i][\delta x\ \delta y\ \delta z\ dx\ dy\ dz]^{\mathrm{T}}}$$

where

$$\begin{array}{lll}
\lambda_1^i = f\mathrm{C1}_i\mathrm{C2}_i, & \lambda_2^i = f((f-\mathrm{C3}_i)\mathrm{C3}_i - \mathrm{C1}_i^2), & \lambda_3^i = f\mathrm{C2}_i(\mathrm{C3}_i - f), \\
\lambda_4^i = f(f-\mathrm{C3}_i), & \lambda_6^i = f\mathrm{C1}_i, & \lambda_8^i = f(\mathrm{C2}_i^2) - (f-\mathrm{C3}_i)\mathrm{C3}_i, \\
\lambda_9^i = -f\mathrm{C1}_i\mathrm{C2}, & \lambda_{10}^i = f\mathrm{C1}_i(f-\mathrm{C3}_i), & \lambda_{12}^i = f(f-\mathrm{C3}_i), \\
\lambda_{13}^i = f\mathrm{C2}_i, & \lambda_{15}^i = -\mathrm{C2}_i(f-\mathrm{C3}_i), & \lambda_{16}^i = \mathrm{C1}_i(f-\mathrm{C3})_i), \\
\lambda_{20} = f - \mathrm{C3}_i, & \lambda_{21}^i = (f-\mathrm{C3}_i)^2, &
\end{array}$$

$$\begin{aligned}
\mathrm{C1}_i &= r_{11}x_i + r_{12}y_i + r_{13}z_i + t, \\
\mathrm{C2}_i &= r_{21}x_i + r_{22}y_i + r_{23}z_i + t_y, \text{ and} \\
\mathrm{C3}_i &= r_{21}x_i + r_{22}y_i + r_{23}z_i + t. \\
i &= 1, 2
\end{aligned}$$

where r_{ij} and t_k are parameters that depend on the orientation and translation parameters of the sensor setting.

The displacement error in the dimension of the line segment consists of two components, the horizontal component ε_{dx} and the vertical component ε_{dy}. The horizontal component, ε_{dx}, equals to $\varepsilon_{d_{u1}} - \varepsilon_{d_{u2}}$, and similarly, the vertical component, ε_{dy}, equals to $\varepsilon_{d_{v1}} - \varepsilon_{d_{v2}}$. The probability density function of ε_{dx} and ε_{dy} can be obtained by integrating the probability density functions of $\varepsilon_{d_{u1}}$ and $\varepsilon_{d_{u2}}$, as well as, $\varepsilon_{d_{v1}}$ and $\varepsilon_{d_{v2}}$, respectively.

$$f_{\varepsilon_{dx}}(\varepsilon_{dx}) = \int_{-\infty}^{\infty} f_{\varepsilon_{d_{u1}}}(\varepsilon_{dx} + \tau) f_{\varepsilon_{d_{u2}}}(\tau)\, d\tau\,, \text{ and}$$

$$f_{\varepsilon_{dy}}(\varepsilon_{dy}) = \int_{-\infty}^{\infty} f_{\varepsilon_{d_{v1}}}(\varepsilon_{dy} + \tau) f_{\varepsilon_{d_{v2}}}(\tau)\, d\tau\,.$$

Similar to the geometric approximation used for the spatial quantization error, the two-dimensional displacement error, ε_d, in the dimension of a linear segment with an angle of γ between the segment and the horizontal axis of the image can be expressed in terms of the horizontal and vertical components of the dimensioning errors due to displacement and the angle, i.e.

$$\tilde{\varepsilon}_d = \cos\gamma\ \varepsilon_{dx} + \sin\gamma\ \varepsilon_{dy}.$$

The probability density function of the dimensioning error due to the displacement of the sensor can be expressed in terms of the probability density functions of the component dimensioning error and the angle, γ, in the following open form:

$$f_{\tilde{\varepsilon}_d}(\tilde{\varepsilon}_d) = \frac{1}{|\cos\gamma|}\int_{-\infty}^{\infty} f_{\varepsilon_{dx}}\left(\tan\gamma\left(\frac{1}{\sin\gamma}\tilde{\varepsilon}_d - \tau\right)\right) f_{\varepsilon_{dy}}(\tau)\, d\tau.$$

3.4. *Integration of quantization errors and displacement errors*

Given a sensor setting and the properties of the sensor, the distribution of the errors in dimensioning need to be determined in order to assess the accuracy of the measurement. To make this assessment, an integration of all significant errors in the active vision inspection is necessary. The spatial quantization errors depend on the resolution of the active sensor. The displacement errors result because of the translational and orientational errors of the active head. Displacement is independent of the resolution of the sensor and, consequently, independent of the quantization error. Thus, the total inspection error, ε_i, is the sum of the quantization error, ε_q, and the displacement error, ε_d, i.e.

$$\varepsilon_i = \varepsilon_q + \varepsilon_d.$$

The probability density function of the total error is computed from the convolution of the pdfs of the quantization and displacement errors as shown in the following equation:

$$f_{\varepsilon_i}(\varepsilon_i) = \int_{-\infty}^{\infty} f_{\varepsilon_q}(\varepsilon_i - \tau) f_{\varepsilon_d}(\tau)\, d\tau.$$

3.5. *Planning accuracy in inspection of linear dimensions*

Accuracy in active vision inspection can be improved by the careful choice of the sensor settings for each inspected dimension. A sensor setting determines the location and view direction where an active sensor may be placed to observe one or more objects which contain one or more topologic entities whose dimensions are to be measured. A sensor arrangement, $\mathrm{S} = \{x_1, x_2, \ldots, x_n\}$, has n sensor settings from x_1 to x_n. x_i is a sensor setting which is an ordered triple (v_i, d_i, o_i) consisting of a sensor location v_i, a sensor view direction d_i, and a set of observable segments (features) o_i from the given setting. The same edge segment of a part (model) can be observed using many sensor settings. Although a dimensional attribute, such as the width of a slot may be observable from different sensor settings, the utility of different sensor settings in dimensional measurement is not identical. However, because more than one entity (feature) can be captured on each image, it is usually desirable to perform several dimensional inspections from a single sensor setting in order to minimize the sensing operations and the data processing. Thus, such simplistic techniques based on an orthogonal direction and a minimal distance are not viable methods in this case. Instead, it is necessary to evaluate the accuracy attainable from the sensor settings, and then ensure that the potential errors in all dimensional measurements from each setting are acceptable for the verification of required (part) tolerances as indicated in the design. Hence, an analysis of expected accuracy of dimensional inspections in terms of the sensor setting parameters and sensor resolution is necessary in this case.

The inspection accuracy in dimensioning a linear segment can be defined as follows:

$$\text{Accuracy} = 1 - \frac{|\varepsilon|}{L},$$

where L is the image (or projected) length of the segment. This representation can be used to analyze the utility of different sensor settings by evaluating the probability for a dimensioning accuracy to be within a particular tolerance. L can be found in terms of the angle between the segment and the direction of the camera axis, β, the translational distances between the camera and the midpoint of the segment, t_x, t_y, t_z (where t_x, t_y are along the horizontal and vertical image plane axes, and t_z is along the sensor view direction), and the model length of the segment in three dimensions, L_w as shown in Fig. 10(a):

$$L(L_w, \beta, t_x, t_y, t_z, f) = \frac{L_w f \sqrt{t_x^2 \cos^2 \beta + ((f + t_z)\sin \beta - t_y \cos \beta)^2}}{(f + t_z)^2 - 1/4 L_w^2 \cos^2 \beta}.$$

When the camera is pointing at the center of the segment (focal axis meets the midpoint), t_x and t_y are both zero. In this case, the above equation can be simplified to be only in terms of β, L_w, f, and the distance d as shown in Fig. 10(b):

$$L(L_w, \beta, d, f) = L_w f \sin \beta \frac{d + f}{(d + f)^2 + 1/4 L_w^2 \cos^2 \beta}.$$

We obtained expressions for the pdf of ε in terms of the sensor and the sensor setting parameters in previous sections. With both L and the error pdf expressed in terms of these parameters, we can compute the likelihood of achieving a certain accuracy level from particular sensor settings. Comparing the variance of accuracy with the tolerance specified, we can determine the dimensional inspection capability for given sensor settings. Thus, given that the accuracy tolerance is $[1 - \mathrm{T}, 1]$ where $0 \leq \mathrm{T} \leq 1$ for a dimension to be inspected, we can compute the likelihood that the achieved accuracy is within this tolerance range by computing $\Pr\{\text{Accuracy} \geq 1 - \mathrm{T}\}$. For example, if the likelihood is greater than a certain threshold, Th where $0 \leq \mathrm{Th} \leq 1$, the sensor setting may be considered acceptable for inspecting the

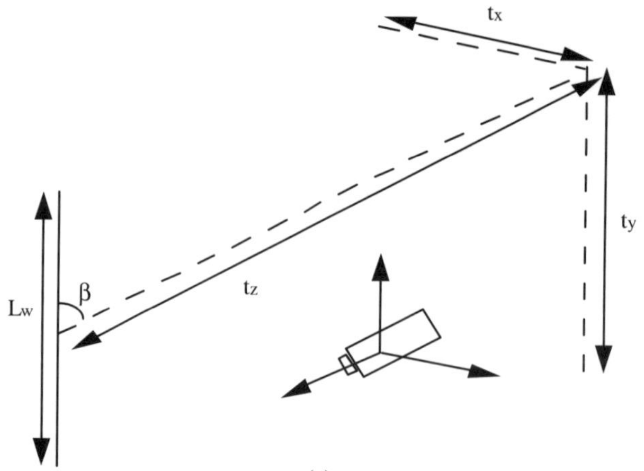

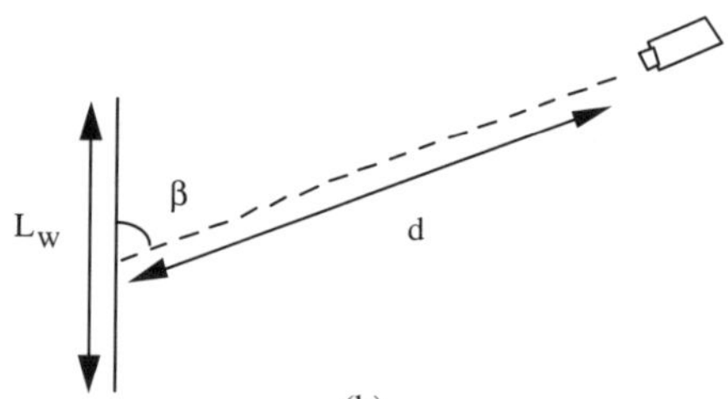

Fig. 10. A line of length L_w is projected onto the camera image plane. (a) The angle between the line and the direction of the camera is β and the translation from the center of the line to the camera in three orthogonal directions is t_x, t_y and t_z. (b) The angle between the line and the direction of the camera is β and the distance from the camera to the center of the line is d.

corresponding dimension, and if the probability is less, the view direction of the sensor and/or the sensor location can be changed. Changing the angle β, and/or the distance d changes the achievable accuracy. Thus, while we had known that a linear segment may be dimensioned from different sensor settings, and that (i) the same segment may have a pixel length twice as long if observed from an orthogonal direction as compared to a 45° angle, (ii) the same segment appears twice as long if observed from a distance D instead of observing it from $2D$, (iii) decreasing the distance D will amplify the error effect of the sensor displacement although the error effect of the spatial quantization will be reduced, and (iv) the imaged segments have different properties with respect to spatial quantization and sensor displacement errors, now that we can probabilistically characterize the integrated effects of these errors. As a result, we can more effectively integrate them in our dimensional inspection strategy, and better understand their quantitative nature. This could be helpful in determining sensor settings which are better suited to dimensionally inspect part (model) attributes with a desired accuracy.

4. Conclusions

In this chapter, we have discussed two inter-related problems in automated visual inspection, (i) the representation of three-dimensional parts and (ii) the planning accuracy based on quantization errors and displacement errors. The Breps are employed to represent the topological and geometric representation of a three-dimensional model; while the entity-based aspect graphs are employed to represent the characteristic views. Applying the geometric reasoning, the features are retrieved and the measurable entities are identified. Based on the information of the entity-based aspect graph, one can determine the potential sensor settings. To ensure the planning accuracy of inspection, the quantization errors and displacement errors are explored, probability density functions of both errors are derived and integrated, and methodology is introduced to determine whether the accuracy of the planned sensor settings is acceptable.

References

1. S. D. Blostein and T. S. Huang, Error analysis in stereo determination of 3D point positions, *IEEE Trans. Pattern Analysis and Machine Intelligence* **9**, 6 (1987).
2. K. Bowyer, Why aspect graphs are not (yet) practical for computer vision, *IEEE Workshop on Directions in Automated CAD-based Vision*, Maui, Hawaii, 1991.
3. K. Bowyer, M. Y. Sallam, D. W. Eggert and J. S. Stewman, Computing the generalized aspect graph for objects with moving parts, *IEEE Trans. Pattern Analysis and Machine Intelligence* **15**, 6 (1993).
4. S. Bryson, Measurement and calibration of static distortion of position data from 3D trackers, *Proc. SPIE Vol. 1669 Stereoscopic Displays and Applications III* (1992).
5. K. Chan and P. Gu, A coordinate measuring machine inspection task planner, *ASME Proc. Manufacturing Science and Engineering*, PED **68-1**, Chicago, illinois (1994).
6. J. Chen and L. M. Chao, Positioning error analysis for robot manipulators with all rotary joints, *Proc. IEEE Int. Conf. Robotics and Automation*, San Francisco, CA (1986).
7. F. W. Ciarallo, C. C. Yang and M. M. Marefat, Displacement errors in active visual inspection, *Proc. IEEE Int. Conf. Robotics and Automation*, Minneapolis, Minnesota (1996).
8. K. Crosby, C. C. Yang, M. M. Marefat and F. W. Ciarallo, Camera settings for dimensional inspection using displacement and quantization errors, *Proc. IEEE Int. Con. Robotics and Automation*, Albuquerque, New Mexico (1997).
9. H. A. Elmaraghy and P. Gu, Expert system for inspection planning, *Annals of the CIRP* **36**, 1 (1987) 85–89.
10. D. W. Eggert, K. W. Bowyer, C. R. Dyer, H. I. Christensen and D. B. Goldgof, The scale space aspect graph, *IEEE Trans. Pattern Analysis and Machine Intelligence* **15**, 11 (1993).
11. H. Freeman, The use of characteristic-view classes for 3D object recognition, *Machine Vision for Three-dimensional Scenes* (Academic Press Inc., 1990) 109–163.
12. P. M. Griffin and J. R. Villalobos, Process capability of automated visual inspection system, *IEEE Trans. Systems, Man, and Cybernetics* **22**, 3 (1992).
13. C. Ho, Precision of digital vision systems, *IEEE Trans. Pattern Analysis and Machine Intelligence* **5**, 6 (1983).

14. C. H. Hoffmann, *Geometric and Solid Modeling: An Introduction* (Morgan Kaufmann Publisher, 1989).
15. S. A. Hutchinson, R. L. Cromwell and A. C. Kak, Planning sensing strategies in a robot work cell with multi-sensor capabilities, *IEEE Trans. Pattern Analysis and Machine Intelligence* **5**, 6 (1986).
16. B. Kamgar-Parsi, Evaluation of quantization error in computer vision, *IEEE Trans. Pattern Analysis and Machine Intelligence* **11**, 9 (1989).
17. J. J. Koenderink and A. J. van Doorn, The internal representation of solid shape with respect to Visio, *Biological Cybernetics* **32** (1979).
18. A. Laurentini, Introducing the reduced aspect graph, *Pattern Recognition Letters* **16** (1995) 43–48.
19. C. Menq, H. Yau and G. Lai, Automated precision measurement of surface profile in CAD-directed inspection, *IEEE Trans. Robotics and Automation* **8**, 2 (1992).
20. C. Menq, C. L. Wong and H. Yau, An intelligent planning environment for automated dimensional inspection of manufactured objects, *ASME Proc. Symp. Concurrent Product Design*, San Francisco (1989).
21. C. Menq and J. Borm, Statistical measure and characterization of robot errors, *Proc. IEEE Int. Conf. Robotics and Automation* (1988).
22. F. L. Merat, G. M. Radack, K. Roumina and S. Ruegsegger, Automated inspection planning within the rapid design system, *Proc. IEEE Int. Conf. Systems Engineering*, Fairborn, OH (1991).
23. F. L. Merat and G. M. Radack, Automatic inspection planning within a feature-based CAD system, *Robotics and Computer-Integrated Manufacturing* **9**, 1 (1992) 61–69.
24. H. D. Park, *CAVIS: CAD Based Automated Visual Inspection System*, PhD Dissertation, Purdue University, 1988.
25. H. K. Park and O. R. Mitchell, CAD based planning and execution of inspection, *Proc. IEEE Computer Vision and Pattern Recognition Conference*, Ann Arbor, Michigan (1988).
26. G. M. Radack and F. L. Merat, The integration of inspection into the CIM Environment, *Proc. Hawaii Int. Conf. System Science* (1990).
27. J. Renders, E. Rossignol, M. Becquet and R. Hanus, Kinematics calibration and geometrical parameter identification for robots, *IEEE Trans. Robotics and Automation* **7**, 6 (1991).
28. R. C. Smith and P. Chessman, On the representation and estimation of spatial uncertainty, *Int. J. Robotics Research* **5**, 3 (1986).
29. S. Su and C. S. G. Lee, Manipulation and propagation of uncertainty and verification of application of actions in assembly tasks, *IEEE Trans. Systems, Man, and Cybernetics* **22**, 6 (1992).
30. W. K. Veitschegger and C. Wu, A method for calibrating and compensating robot kinematic errors, *Proc. IEEE Int. Conf. Robotics and Automation* (1987).
31. C. C. Yang, M. M. Marefat and F. W. Ciarallo, Analysis of errors and planning accuracy on dimensional measurement in active vision inspection, *IEEE Trans. Robotics and Automation*, **14**, 3 (1998) 476–487.
32. C. C. Yang, M. M. Marefat and E. J. Johnson, Entity-based aspect graph: making viewer centered representations more efficient, *Pattern Recognition Letters* **19**, 3–4 (1998) 265–277.
33. C. C. Yang and M. M. Marefat, Object-oriented concepts and mechanisms for featured-based computer integrated inspection, *Advances in Engineering Software* **20**, 2/3 (1994) 157–179.

34. C. C. Yang and F. W. Ciarallo, Minimizing the probabilistic magnitude of active vision errors using genetic algorithm, *Proc. IEEE Int. Con. Systems, Man, and Cybernetics*, Orlando, Florida (1997).
35. C. C. Yang, M. M. Marefat and F. W. Ciarallo, Analysis of errors in dimensional inspection based on active vision, *Proc. SPIE Int. Symposium on Intelligent Robots and Computer Vision XIII: 3D Vision, Product Inspection, Active Vision*, Cambridge, MA (1994).
36. C. C. Yang and M. M. Marefat, Spatial quantization error in active vision inspection, *Proc. IEEE Int. Con. Systems, Man, and Cybernetics*, San Antonio, TX (1994).
37. C. C. Yang, M. M. Marefat and R. L. Kashyap, Automated visual inspection based on CAD models, *Proc. IEEE Int. Con. Robotics and Automation*, San Diego, CA (1994).
38. C. C. Yang and M. M. Marefat, Flexible active computer integrated Visual Inspection, *Proc. NSF Design and Manufacturing Conference*, Cambridge, MA (1994).
39. H. Yau and C. Menq, Path planning for automated dimensional inspection using coordinate measuring machines, *Proc. IEEE Int. Conf. Robotics and Automation*, Sacramento (1991).

INDEX